Research Reports ESPRIT

Project 7096 · CCE-CNMA · Volume 2

Edited in cooperation with the European Commission

Springer
Berlin
Heidelberg
New York
Barcelona
Budapest
Hong Kong
London
Milan
Paris
Tokyo

ESPRIT Consortium CCE-CNMA (Eds.)

MMS: A Communication Language for Manufacturing

Springer

Volume Editors

ESPRIT Consortium CCE-CNMA
c/o CSC Manufacturing Practice
Guild Centre Offices
Lords Walk, Preston, Lancashire PR1 1RE, UK

ESPRIT Project 7096, CCE-CNMA (CIME Computing Environment: Integrating CNMA) belongs to the CIME sector (Computer-Integrated Manufacturing and Engineering) of the ESPRIT Programme (European Specific Programme for Research and Development in Information Technologies) supported by the European Commission.
CCE-CNMA builds on the results of ESPRIT projects 955, 2617, and 5104 on CNMA (Communications Network for Manufacturing Applications), progressing to new areas of technology.
The main objective of the project is to specify, implement and validate an environment for the development and execution of manufacturing applications. This environment integrates technologies and standards from various organisations such as OSF (Open Software Foundation), ISO or X/OPEN. The consortium continues work on a communication infrastructure for industrial applications that is based on international standards and develops tools for migration to and administration of this vendor-independent communication network.
The 16 members of the ESPRIT consortium CCE-CNMA consist of CIM Users, CIM Vendors, System Integrators and Academic Institutes from 6 European countries.

Cataloging-in-Publication Data applied for

Die Deutsche Bibliothek - CIP-Einheitsaufnahme

MMS: a communication language for manufacturing / ESPRIT Consortium CCE-CNMA (ed.). - Berlin ; Heidelberg ; New York : Springer, 1995
(Research reports ESPRIT : Project 7096, CCE-CNMA ; Vol. 2)
ISBN 3-540-59061-7
NE: ESPRIT Consortium CCE-CNMA; Research reports ESPRIT / 7096

CR Subject Classification (1991): C.2, J.6, D.3-4, J.1, H.2
ISBN-13: 978-3-540-59061-3 e-ISBN-13: 978-3-642-79639-5
DOI: 10.1007/978-3-642-79639-5

Publication No. EUR 16172 EN of the European Commission, Dissemination of Scientific and Technical Knowledge Unit, Directorate-General Telecommunications, Information Market and Exploitation of Research, Luxembourg.

Typesetting: Camera-ready by the editors
SPIN: 10486046 45/3142-543210 – Printed on acid-free paper

Preface

Today's manufacturing environment is characterised by significant change in the way it is able to respond to its business objectives. Companies now face new challenges to meet customer demands including greater flexibility, a wider range of products, higher quality, improved lead time, whilst maintaining competitive prices.

The changes manifest in each company will differ but, the trend is towards providing a leaner, more responsive organisation. This is reflected in the need for Information Technology (IT) systems to be more integrated across an enterprise and for the systems to be provided on a distributed basis. At the same time it is imperative that the costs associated with these IT systems and the range of skills required by an enterprise to develop and support such systems should be minimal.

One way of achieving this objective is to adopt the use of standards. During recent years there has been a lot of activity to standardise the way that manufacturing devices communicate with each other. Considerable progress has been made, thanks not least to the success of the international effort of standardisation bodies like ISO, IEEE and IEC, and projects applying standardised protocols like Manufacturing Automation Protocol (MAP), in the USA, Communications Network for Manufacturing Applications (CNMA) in Europe and not forgetting Mini-MAP and Fieldbus technology.

The most notable standard defined is ISO 9506-1 and 9506-2, Manufacturing Message Specification (MMS) which provides the generic services designed to facilitate the interworking of programmable devices on the shop floor. These services are intended to be enhanced and augmented by companion standards, each of which is oriented towards a specific class of application.

The purpose of this book is to describe what MMS is and how it can be used, taking for example a real production facility at Mercedes-Benz, Germany. This book is intended for engineers who are responsible for the specification, design, development and integration of manufacturing systems. Familiarity with manufacturing applications, existing communication protocols, and how manufacturing equipment and devices can be integrated will increase the reader's appreciation of the contents. The reader of this book will gain an understanding of how MMS can be used, to derive its benefits and subsequently be able to either specify their own requirements, or assess the advantages of solutions provided by vendors or systems integrators.

References at the end of this book point to additional reading.

May 1995

Lawrence Cox / Patrick Pleinevaux

♣ ♣ ♣ ♣ ♣ ♣

Trademarks

The following trademarks used in this book are hereby acknowledged:

- *MAP* is a registered trademark of *General Motors*.
- *Intel* is a registered trademark of *Intel Corporation*
- *Microsoft, Windows, MS, DOS, Visual Basic, Visual C++, ODBC, Access, Excel* are registered trademarks of *Microsoft Corporation*
- *IBM, SNA, CICS, CPI-C* are registered trademarks of *International Business Machines*
- *INGRES* is a registered trademark of *Ingres Corp.*
- *SINEC H1, SINEC AP-TF* are registered trademarks of *Siemens*
- *UNITELWAY, MAPWAY* are registered trademarks of *Telemecanique*
- *MODBUS* is a registered trademark of *MODICON*
- *DECnet* is a registered trademark of *Digital*
- *Lotus 1-2-3* is a registered trademark of *Lotus Development Corporation*
- *ORACLE* is a registered trademark of *Oracle Corporation*

Contents

1. Introduction

This chapter discusses the need for standards in the context of manufacturing, highlighting the most significant initiatives to date. It also discusses the motive for producing the book, a summary of each chapter and acknowledges the work of the authors.

1.1 The need for standards

Today's manufacturing environments operate in a heterogeneous manner, characterised by a multitude of different methods, applications and technologies. Customer demand in the way of greater product choice, higher quality, quicker delivery times and lower costs is forcing companies to create a leaner, more responsive organisation. This in turn is mandating Information Technology (IT) systems become more integrated across an enterprise and for the systems to be provided on a distributed basis.

However, the task of integrating the systems in an economical manner has become more apparent since investment in capital and development must be justified. Manufacturing applications can be complex and can require different levels of investment in established state-of-the-art technology. Moreover, in order to support process chains in the technical product life cycle, a close link to the engineering applications, e.g. concerned with product design, is required. Since many of these applications are computerised, the requirement to automate their interfaces with one another becomes more of a priority. Therefore by providing a Computer Integrated Manufacturing and Engineering (CIME) environment within the organisation, an essential gap is filled to complement the overall business objectives.

CIME and Factory Automation are characterised by the large variety of computers and controllers from many vendors in industrial installations supporting numerous application domains. During integration, the industrial user is faced with two options. One option is to purchase from a single vendor and to use that vendor's proprietary solutions only to achieve an integrated environment. This option ties the user to that vendor's products and is therefore restrictive. His choice is then limited for devices and computers in the future. The second option is to integrate computers and controllers from more than one vendor. However, this option forces the user into considerable expenditure to achieve communic-ation and cooperation between numerous proprietary systems.

Thus, users who have experienced these problems primarily have a requirement for a single infrastructure supported by all possible vendors. To achieve this, evolution towards Computer Integrated Manufacturing must be considered, ensuring that common standards are adopted by both vendors and users. This is especially true in the key areas of industrial Local Area Network communications and data management. In recent years, work has been carried out under the auspices of the International Standard Organisation (ISO)'s "7-Layer Model" for Open Systems Interconnection (OSI). The most publicised examples of this work in the manufacturing domain have been the General Motors MAP (Manufact-uring Automation Protocol) and the Boeing TOP (Technical Office Protocol) init-iatives in the USA.

A complementary European initiative was started in January 1986, leading to the formation of CNMA (Communications Network for Manufacturing Appl-ications), a multi-national collaborative ESPRIT project to address the needs of industry in the European context. The work of this project has expanded upon, but still includes its original ambitions to find a practical and realistic solution for industry.

The work of the IROFA FAIS (Factory Automation Interconnection System) project, based in Japan has resulted in the development of a new Mini-MAP 1993 version.

There are many other initiatives which have been started throughout the world as a result of these activities, all attempting to exploit the work of open system developments.

Whilst the application layer of the MAP profile has been widened from MMS, FTAM, Network Management and Directory Services to also optionally include Remote Database Access, X-Windows, and Virtual Terminal protocols to support typical applications in manufacturing, in MAP 3.0 1993 the main application protocol MMS has now been made ***mandatory***.

Whilst this condition represents a significant demonstration of the value of the MMS service, it is the belief of the CNMA project that further work is required to popularise its value amongst the CIM community. In the next section we describe the motive for publishing this book.

1.2 Motivation for the book

The motivation for this book resulted from concern about the uptake of MMS through vendors products, in particular vendor partners participating in the ESPRIT project CCE-CNMA EP7096, which has been the primary research project investigating the suitability of MMS across a wide range of manufacturing sectors. It was felt that increased awareness and familiarity with the service would highlight the significant business benefits to be gained.

1.3 Purpose of the book

The purpose of this book is to describe what MMS is and how it can be used, taking for example a real production facility at Mercedes-Benz, Germany. This book is intended for engineers who are responsible for the specification, design, development and integration of manufacturing systems. Familiarity with manufacturing applications, existing communication protocols, and how manufacturing equipment and devices can be integrated will increase the reader's appreciation of the contents. The reader of this book will gain an understanding of how MMS can be used, to derive its benefits and subsequently be able to either specify his own requirements, or assess the advantages of solutions provided by vendors or systems integrators.

1.4 Organisation of the book

This book is intended to be used as a reference manual but, for the reader who is less experienced in MMS concepts, Chapters 2 and 3 should be read first. A brief summary of each chapter follows:

Chapter 2, describes the '*Rationale and Context*' for introduction of MMS and presents the major communication networks that use this protocol. An indication of the benefits that the user can expect in using these networks is also provided.

Chapter 3, '*Manufacturing Message Specification (MMS)*' introduces the basic concepts that underlie MMS: the client/server model, the object modelling approach and the standards related to MMS.

Chapter 4, presents a description of the '*Most Important MMS Object Classes*' that are implemented by the majority of device manufacturers. These are the VMD, Domain, Program Invocation and Named Variable objects. Great importance is attached to the practical use of MMS which is reflected in numerous examples. These examples should help the reader understand the basis of MMS.

Chapter 5, '*MMS in Real Manufacturing Devices* ' analyses three implementations of MMS for manufacturing devices from Siemens and Télémécanique. An NC is used to illustrate how application-specific objects can be modelled using MMS objects and with which MMS services they can be accessed.

Chapter 6, describes the '*Use of MMS in a Real Manufacturing Application*', within a modern production facility at Mercedes-Benz, used for the manufacture of car engines. The facility consists of machining centres, workstations and storage facility serviced by a rail guided vehicle. The example scenarios are considered helpful to the general understanding of MMS in a wide range of typical manufacturing applications.

Chapter 7, presents '*MMSI*', the application programming interface specified by General Motors for MMS. A number of examples are included to give the reader an idea of this interface. The benefits of using this interface are discussed.

Chapter 8, describes '*A DDE Interface to MMS*', that is available in the Windows environment. The MMS/DDE server, which has been developed in the CCE-CNMA project allows direct access to manufacturing devices on the shopfloor via MMS. A practical example is given at the end of the chapter to illustrate the use of this server from an Excel spreadsheet.

Chapter 9, introduces '*Other MMS Objects and Services*' that are sometimes implemented; these are concerned with operator communication, semaphores, events and journals.

Chapter 10, '*Conclusion*' provides a summary of what has been discussed, and then concentrates on the importance of MMS, its multi-sectorial applicability, the real benefits to the user, interoperability, where to obtain more information on CNMA and MMS and, finally, a perspective of the future developments.

Appendix 1, '*The MMS and CNMA Information Servers*' provides details on how to access further information on MMS and CNMA using an information server on World Wide Web.

Appendix 2, provides a list of '*MMSI Context Sensitive Functions*' for each MMS functional unit, and a short description of the function. This detail is an extension of what is described in Chapter 7.

1.5 Acknowledgements

The ESPRIT CCE-CNMA consortium represents 17 organisations from 6 European countries, a large project when it is compared with other initiatives within the ESPRIT framework. This has been possible due to the special synergy which exists between the partners as a result of the successes of CNMA. We have to thank the European Commission for providing such a framework for this co-operative research.

As in previous phases of CNMA, this consortium has provided the appropriate skills and resources to present the many different perspectives necessary to deal with the problems of manufacturing. The partners listed below provide a well balanced force to deal with the demands of today's manufacturing environment.

Users	**Systems Integrators**
British Aerospace	Alcatel-TITN-Answare
Aerospatiale	Fraunhofer IITB
EFACEC	Silicomp Ingenierie
Magneti Marelli	O. Group (part of Olivetti)
Mercedes-Benz	Robotiker

Vendors	**Academic Institutes**
Bull	EPFL-LIT
Olivetti	University of Porto
Siemens	University of Stuttgart
Siemens-Nixdorf	

We also wish to acknowledge the contribution made in previous phases of this project by our earlier partners: *Aeritalia (I), BMW (D), Elf Aquitaine (F) and PSA (F) from ESPRIT project 955 (CNMA) and CEGELEC Projects Ltd (UK), Com-Consult Communication Technologies (D) and Renault (F) from ESPRIT project 2617 (CNMA).*

The consortium acknowledges the work of the authors whose experience in implementing OSI solutions has helped to make this book unique. Their area of contribution is identified in the book and we list their affiliations below.

Authors	**Affiliation**
Jean-Pierre Fayolle	Bull, Grenoble, France.
Lawrence Cox	CSC Manufacturing Practice (for British Aerospace) Preston, England.
Patrick Pleinevaux	EPFL-LIT Lausanne, Switzerland
Werner Blumenstock Artur Lederhofer Frank Sackmann Thomas Talanis	Siemens AG Nuernberg-Moorenbrunn, Germany
Manfred Gaertner	University of Stuttgart, Germany

Special thanks are due to Werner Blumenstock (Siemens), Miguel Melchor (Robotiker), Cataldo Nichilo (Olivetti Ricerca), Ulli Fischer (Mercedes-Benz) and Raul Oliveira (University of Porto) for their work in reviewing this manuscript.

The final editors are Patrick Pleinevaux (EPFL-LIT), Technical Manager for CCE-CNMA and Lawrence Cox (CSC Manufacturing Practice) Project Coordinator for CCE-CNMA, representing British Aerospace.

Further acknowledgement must be made of the European Commission's special interest in providing opportunities for promotion of research and development activities, which has helped realise the publication of this book. The standard publishing arrangements between the EC and Springer-Verlag, Heidelberg have greatly simplified the technical, commercial and copyright issues that would otherwise have been experienced if using a normal publishing agreement. The EC's strategy to support information dissemination activities for EC partly funded research activities, like the CCE-CNMA project is widely appreciated within this consortium.

The final contribution we present here, has been fully approved by this consortium.

May 1995

Lawrence Cox
on behalf of
ESPRIT Consortium CCE-CNMA

2. Rationale and Context

This chapter describes the context in which MMS was defined, the rationale for its introduction and gives the position of MMS in the hierarchy of communication protocols. The MAP, Mini-MAP and Profibus networks are then briefly presented. We indicate the benefits that the user can get in using these networks.

2.1 Rationale for MMS

The problem discussed in this section is typical of many different application domains, not just manufacturing. In the industrial sector, process control but also building automation and the food industry have similar problems.

Let us consider a manufacturing cell. In this cell, Numerical Controllers are supplied by vendor A, Programmable Logic Controllers by vendor B and the cell controller by vendor C. The designer of the cell is faced with a fundamental problem: heterogeneity of the devices, due to the variety of suppliers. The consequence of this problem is that the communication languages - protocols - spoken by the devices are all different. Devices of vendor A speak language L_A while devices of vendor B talk L_B and the devices are unable to talk to one another. Engineers responsible for the cell control application have two options:

1. Use languages L_A and L_B in their programs to communicate with devices A and B.
2. Use a common language L_M independent of A and B but that A and B understand. This language is MMS.

The problems with the first option are:

- Engineers need to learn at least two different communication languages.
- The program is dependent on languages L_A and L_B and must be modified if devices from another supplier D are used.
- Programs are more complex and more expensive to maintain.

MMS is a communication language, specifically designed for industrial applications. It is vendor-independent, standardised by ISO and available on a very wide range of industrial devices. All major device manufacturers have implemented MMS on their devices.

2.2 The OSI reference model

Many communication protocols have been designed and implemented over the last twenty years. These protocols solve different problems and are generally classified according to the level at which they operate. In this section, we clarify the position of MMS in the hierarchy of communication protocols.

Communication between computer systems is a complex task. In this chapter, communication means the activity that is concerned with exchanging messages between machines, independently of their location in the factory, the length of messages, the communication problems, etc. A **communication network** is a system that delivers a **service** that allows users to transfer messages between them. This is represented in Figure 2.1.

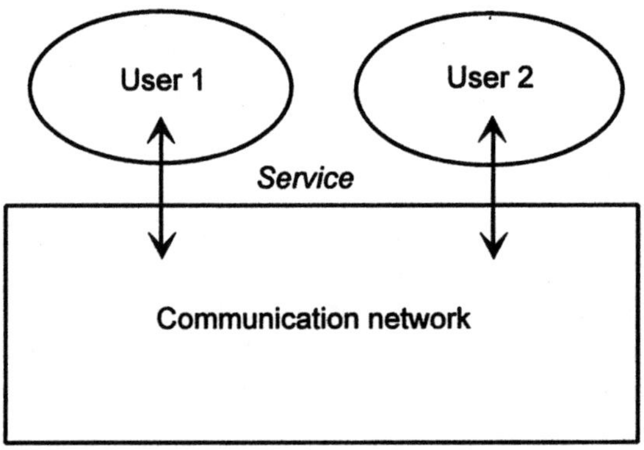

Figure 2.1: A communication network as service provider

This service is implemented by a set of protocols that are followed by the different machines involved in the communication. A **protocol** can be seen as a set of rules that govern how and when messages are sent, as a description of the messages that can be exchanged and a set of procedures to build these messages.

ISO have proposed a model to describe the different functions involved in a communication. This model, known as the OSI (Open Systems Interconnection) Reference Model, identifies seven levels of functions called layers as shown in Figure 2.2.

These layers can be classified into two groups:

1. Transfer layers - layers 1 to 4 - deal with the problem of transferring a raw message (a sequence of octets) from one machine to another, whatever its location in the world and the problem that can affect this transfer. The network layer for example is responsible for routing messages between subnetworks of the whole network. The transport layer on the other hand, is responsible for detecting errors and recovering from such errors if possible.
2. Application-oriented layers are concerned with providing the expected service to the user program and are not concerned with transfer problems. The session layer provides synchronisation mechanisms between communicating partners. The presentation layer deals with data representation problems, hiding the differences in data format and preserving the meaning of data.

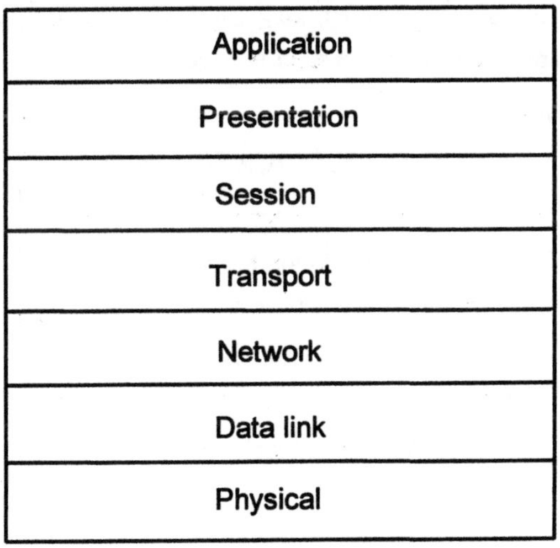

Figure 2.2: OSI Reference Model

The application layer, just below the user program, can be seen as offering a language composed of a set of services that can be invoked by the user to remotely control a device, whether it be a terminal, a file store or a manufacturing device.

> *MMS is an application layer protocol, offering a rich language that can be used to communicate with industrial devices.*

2.3. MAP and CNMA

Depending on the problem to solve, different communication networks can be defined. For almost 20 years, computer and device manufacturers have defined their own communication protocols and communication networks to talk to their devices. Section 2.1 discussed the typical problem encountered by industrial companies when building manufacturing or process control applications. General Motors, at the beginning of the 1980s, tried to solve this problem by defining its own, vendor independent computer network and called it **MAP** (Manufacturing Automation Protocol). In 1986, a project called **CNMA** (Communications Network for Manufacturing Applications) was started in Europe to define a communication network, compatible with MAP, that satisfies the requirements of the European industry. MAP and CNMA provide essentially the same services and can be used in the same application by different devices. They are both composed of seven layers and the protocols adopted are international standards. The architectures of MAP and CNMA can be seen in Figure 2.3.

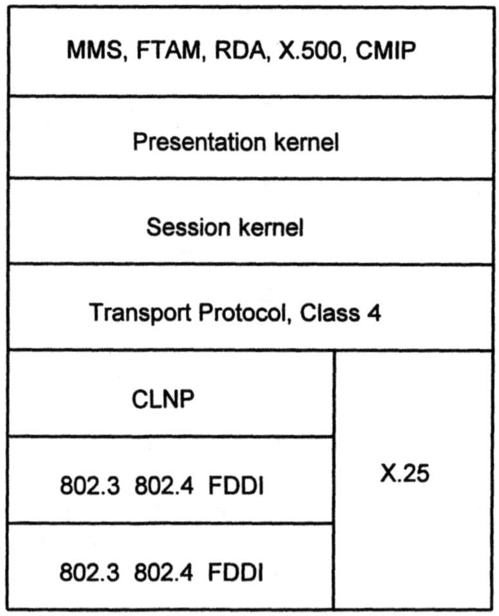

Figure 2.3: MAP and CNMA architectures

Layers 1 to 6 are very general and not specific to manufacturing applications. In layer 7, all protocols except one are general: FTAM for file transfer and management, RDA for database access, X.500 for access to a directory service. The only specific protocol is MMS.

The fact that MMS relies on six other layers has the following benefits for the user:

- Communication can be between two nodes of two different networks, anywhere in the world.
- The user does not have to worry about message length, message losses or duplication.
- Messages are delivered in the order they were sent.
- Differences in data representation are dealt with by the communication network.
- Detection of peer failure is automatically done and notified by the network.

The interested reader will find more information on MAP in [Ciminiera 92] and on CNMA in [CCE 95].

2.4 Mini-MAP

Mini-MAP is another communication network proposed for industrial applications. Contrary to MAP, it relies on a 3-layer architecture where layers 3 to 6 have been omitted. A benefit of this removal is performance. The price to pay for this increased performance is a reduced functionality. The following problems are not dealt with by Mini-MAP and may limit its applicability:

- Communication outside of the local subnetwork: Mini-MAP messages cannot cross bridges and routers.
- Segmentation and reassembly of long messages: the user must check the length of the messages he generates.
- Automatic detection of peer failure.
- Security

Mini-MAP has gained little support from vendors in the world. A Japanese project called FAIS (Factory Automation Interconnection System) attempted to solve some of the problems of Mini-MAP and the improvements found have been included in the 1993 revision of MAP 3.0. This architecture is represented in Figure 2.4. In spite of this, Mini-MAP suffers from a major problem: it relies on the IEEE 802.4 token passing bus protocol at layer 2, a complex and expensive technology.

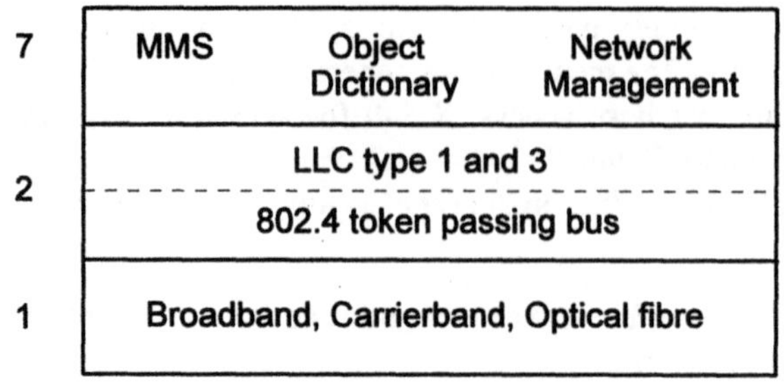

Figure 2.4: Mini-MAP 1993

2.5 Profibus

Profibus is a vendor independent communication network proposed for industrial applications. Designed as a field bus, to interconnect sensors and actuators, it is used also to interconnect manufacturing devices and control computers. As with Mini-MAP, it relies on a 3-layer architecture where layers 3 to 6 have been omitted as can be seen in Figure 2.5. Profibus has the same advantages and limitations as Mini-MAP. A particular advantage of Profibus is cost: it is much cheaper than MAP and Mini-MAP.

Profibus offers MMS-like functionality at the application layer. FMS (Fieldbus Message Specification) is a service and protocol that is very close to MMS. FMS however is simpler than MMS: many services and parameters have been either removed or simplified for example for variable, semaphore and event management.

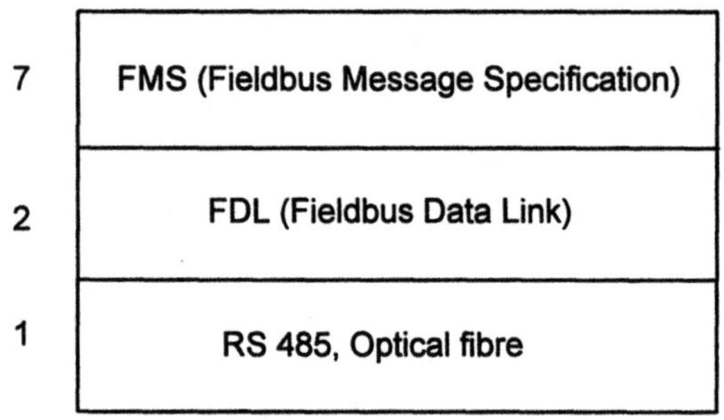

Figure 2.5: Profibus architecture

2.6 Summary

MMS is a communication language designed to interconnect industrial devices in a heterogeneous environment in which devices of multiple suppliers must coexist. It is an application layer service and protocol that offers a rich set of commands to remotely control and monitor industrial devices. MMS is available in MAP and Mini-MAP as defined in the ISO standard and available in a simplified version in Profibus.

3. Manufacturing Message Specification (MMS)

3.1 Basics

MMS is an application layer protocol that offers services for the remote control and monitoring of industrial devices. It can typically be used to communicate with Programmable Logic Controllers (PLCs), Numerical Controllers (NCs) of machines but also Personal Computers and mini computers that act as production controllers. MMS offers a wide range of services to read or write variables, to download or upload programs and data, to remotely control execution of programs, to manage events, etc.

MMS is a multi-part **international standard** that precisely defines:

- a set of objects, that represent resources of the device
- a set of services, that can be used to remotely manipulate these objects
- a set of messages, called PDUs (Protocol Data Units), that carry the requests and responses associated to these services
- a set of automata, indicating which services can be called in which state of the device.

A typical industrial application involving MMS is shown in Figure 3.1. A manufacturing cell is composed of a machining centre, a central store for parts and tools, a transport system and a control computer. The transport system moves parts and tools between the central store and the machines. Co-ordination of the machining, transport and storage operations is performed by a control computer that uses MMS as language for communication with the shop floor devices.

An **MMS user** is an application program which uses the MMS services to communicate with the MMS user of a target system, for example an NC system. For the structuring of the dialogue of users involved in an MMS communication, the standard makes use of the **client/server model**. MMS extends the classical client/server model by allowing the server to request services from a client without changing its role. In the classical client/server model, a server is passive, in that its role is limited to respond to requests from clients. A real MMS implementation may adopt the client role, the server role, or both during the lifetime of an application association. It should be noted that the terms MMS client and MMS server are used to distinguish the two roles an MMS user may adopt.

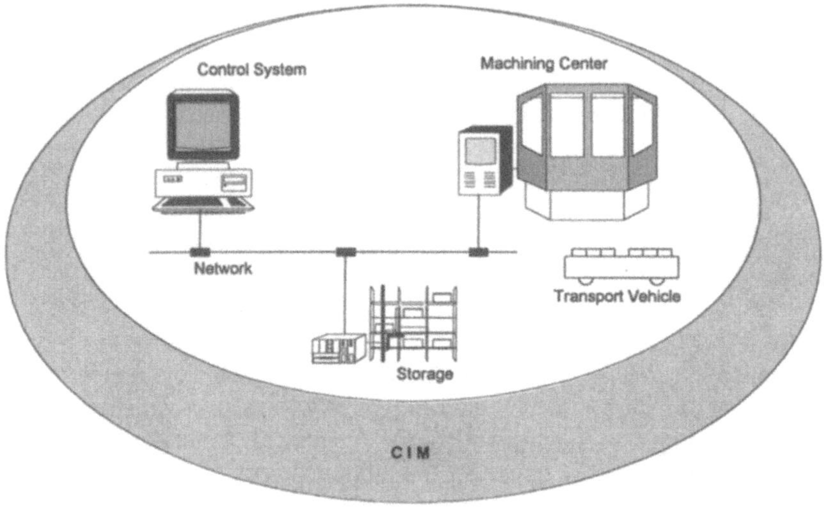

CIM Computer Integrated Manufacturing

Figure 3.1: Typical application area of MMS.

The **MMS provider** is that part of the communication system that provides the user with the MMS services. Essentially, the MMS provider is responsible for transferring MMS requests and responses between MMS users. To do this, the MMS provider relies on the services of the layers below, for example the presentation layer in MAP or the link layer in Mini-MAP. The MMS provider is responsible for building messages - called PDUs - that carry the requests and responses issued by the user and interpreting incoming messages. Figure 3.2 illustrates the notion of MMS provider and the associated concepts of service primitives. A primitive can be seen as a message exchanged between a user and a service provider. Four kinds of primitives are traditionally defined: Request primitives invoked by a user requesting a service, Indication primitives invoked by the provider to convey the request to the remote user; Response primitives, invoked by the remote user to supply his response to the calling user; Confirm primitives, invoked by the provider to convey this response to the requesting user. As can be seen from this definition, the provider acts as a broker between the requesting and the responding users. More importantly, the users have no idea of how the services are implemented.

An **MMS server** or server in the sequel is an application which *maps* onto MMS objects the physical resources which may be accessed from the network. This notion of mapping is fundamental in MMS: clients do not access physical resources directly but the MMS objects that represent them. The standard defines the server "... as the peer communicating entity which behaves as a VMD for a particular service request instance".

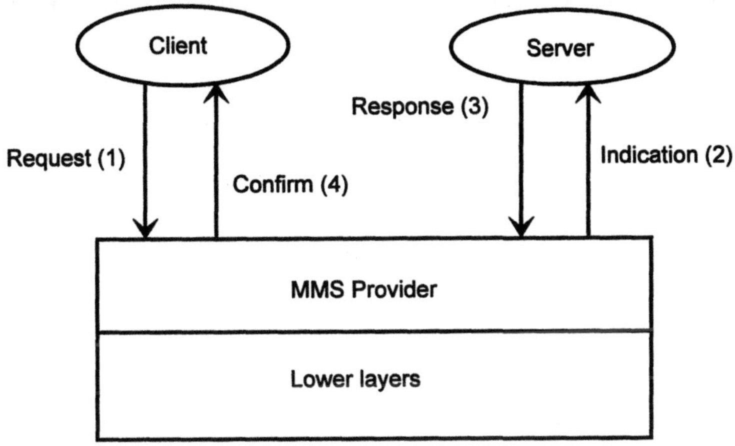

Figure 3.2: MMS Provider and service primitives.

An **MMS client** or client in the sequel is an application which accesses the MMS objects of a server via the network and thus makes use of the available physical resources provided by this MMS server.

3.2 The MMS object model

Object modelling is a technique that has gained wide acceptance for the analysis and design of complex systems. Essentially, an **object** is a representation of a real world entity. An object is characterised by the attributes and operations that allow to access the object attributes. These store information concerning the object. The operations are the only means to access these attributes. The object acts as a server offering operations accessible by other objects called clients. This client-server relationship is a method to structure communication between objects. Figure 3.3 shows the classical object model.

In more sophisticated object models, objects belong to **classes** that are collections of objects sharing the same definition in terms of attributes, operations and behaviour. The members of a class are called **instances**. They have the same definition but different values for their attributes.

MMS adopts the object modelling approach for the definition of the services offered by servers. Twelve classes of objects are defined in MMS that are accessible by name. This chapter introduces the most important objects used in existing manufacturing applications. The operations of classical object models are called **services** in MMS. The attributes of MMS objects are accessible by the MMS services only. The behaviour of objects is defined using finite state machines that indicate the state transitions resulting from the reception of service invocations.

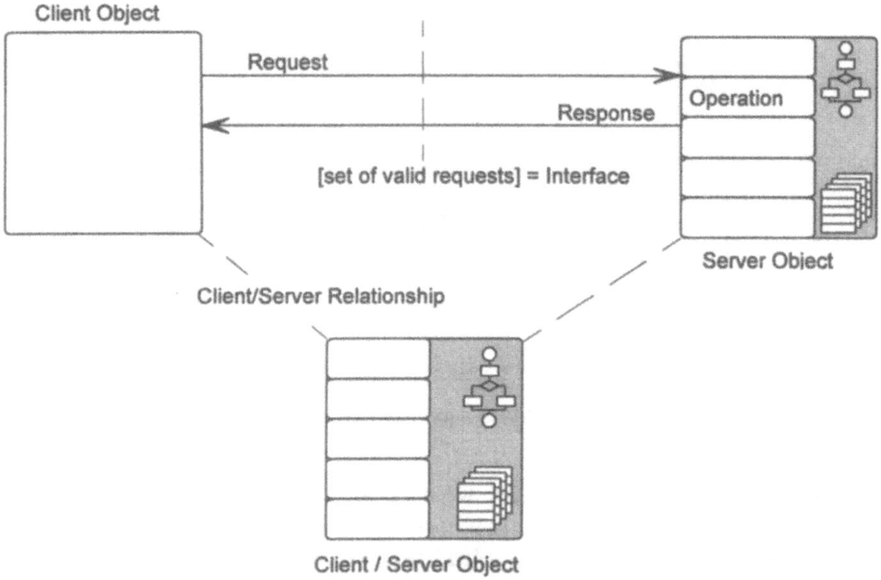

Figure 3.3:The classical object model

Figure 3.4 illustrates the essence of the MMS object modelling approach. The idea is to define device- and application-independent objects that are sufficiently abstract to hide implementations details. The figure displays two planes: in the background, the plane containing real objects, existing in a manufacturing device. These are for example a tool magazine, a tool, a pallet, etc. In the foreground is represented the plane of the MMS objects, which are visible from the application programs and are representations of the real objects behind. The user does not access real objects directly but accesses MMS objects. These are sufficiently general to represent nearly all resources of industrial devices.

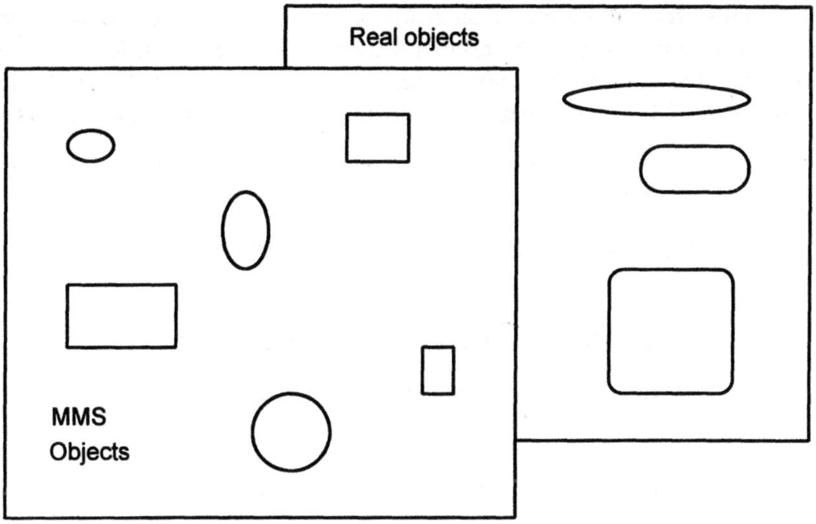

Figure 3.4: MMS object modelling approach

MMS makes use of a template to define the properties of an object class, as shown in Figure 3.5. Each instance of an object class has the same set of attributes, but has its own set of attribute values. The values may be predefined or established by MMS services. An MMS service may change one or more attribute values of one or more instances of an object class.

In object-oriented modelling, each instance of a class must be uniquely identified among all instances of the same class. For this purpose, each class template contains at least one **key attribute** which allows the unambiguous identification of all instances of the class. All MMS object classes have only one key attribute which is in general the name of the object. The name of an object is always bound to a name space or **scope** in which the name must be unique. MMS defines three such scopes: the VMD, Domain and application association. These concepts will be described in 'Description of the Most Important Object Classes' (Chap. 4).

A class template may also contain **conditional attributes** which are relevant if and only if certain conditions are true. A condition constrains the existence of subjected attributes.

There may also be attributes which make reference to instances of other classes in order to reflect links between objects. Such attributes are called **reference attributes** in this book.

Generally, the definition of an MMS object class contains only attributes which are accessible via MMS services. However, in some class definitions there are additional attributes which are not visible to the MMS client. These attributes should help the MMS implementor and cover general implementation aspects.

Object:	name of class		
	Key Attribute:	name of attribute (values)	
	:	:	
	Key Attribute:	name of attribute (values)	
	Attribute:	name of attribute (values)	
	:	:	
	Attribute:	name of attribute (values)	
	Constraint:	constraint expression	
		Attribute:	name of attribute (values)
		:	name of attribute (values)
		Attribute:	name of attribute (values)

Figure 3.5: Template for the definition of objects in MMS

As an example, let us consider the MMS Program Invocation (PI) object, as shown in Figure 3.6. In MMS, a PI is an abstraction of a dynamic element whose execution can be controlled by services such as Start, Stop, etc.

The PI has one key attribute called Program Invocation Name which is a string allowing the identification of the object. The PI has a number of "regular" attributes that describe its properties. The State attribute for example indicates in which state the object is currently.

The List of Domain References is a reference attribute which points to one or more MMS objects that compose the PI. The PI has three conditional attributes which are present only if the Monitor attribute has the value TRUE. These are the Event Condition, Event Action and Event Enrolment reference attributes which point to event objects that are automatically created and initialised by the server when execution monitoring is required by the client.

Object:	Program Invocation		
	Key Attribute:	Program Invocation Name	
	Attribute:	State (IDLE, STARTING, RUNNING, STOPPING, STOPPED, RESUMING, RESETTING, UNRUNNABLE)	
	Attribute:	List Of Domain References	
	Attribute	MMS Deletable	
	Attribute:	Reusable	
	Attribute:	Monitor (TRUE, FALSE)	
	Constraint	Monitor = TRUE	
		Attribute	Event Condition Reference
		Attribute	Event Action Reference
		Attribute	Event Enrolment Reference
	Attribute	Execution Argument	
	Attribute	Additional Detail	

Figure 3.6: The Program Invocation object

3.3. Confirmed and unconfirmed services

In general, MMS services are grouped into two categories: the services which require a confirmation - called confirmed services - and services which do not require such a confirmation - called unconfirmed services. Unconfirmed services can only be invoked by a server, and this, to report a significant event to the client.

The following sections describe the different steps of a communication for confirmed and unconfirmed services. It should be mentioned that these sequences are valid for all confirmed services independently of who requests the service.

3.3.1. Confirmed services

A service which requires a confirmation is called **confirmed service**. The common communication sequence for all confirmed services is shown in Figure 3.7. In this figure, the MMS client requests a confirmed MMS service. Note that MMS defines confirmed services which may be used by the client and confirmed services which may be used by the server. Examples of confirmed services invoked by the server will be given in 'Description of the Most Important MMS Object Classes' (Chapter 4) when presenting the domain download services.

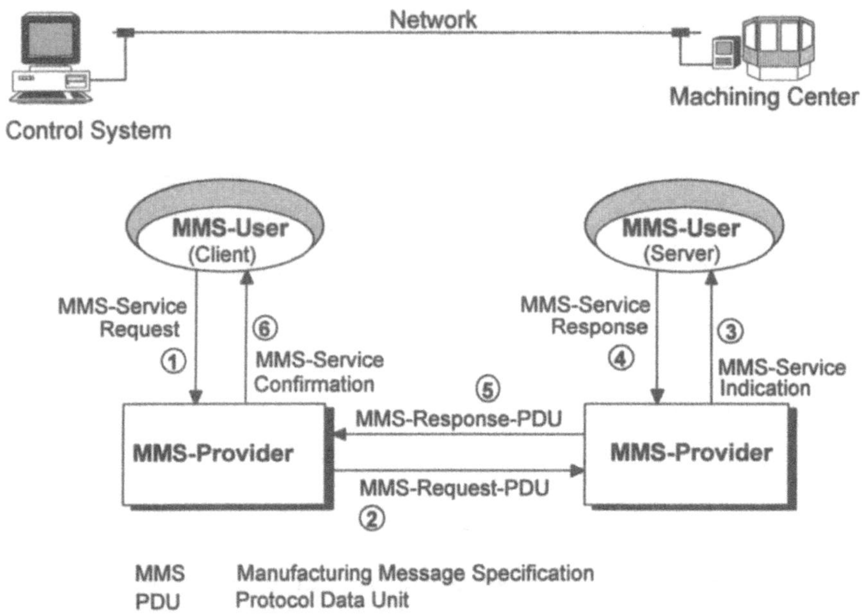

Figure 3.7: Logical Communication Sequence (Confirmed Services)

Table 3.1 details the different steps that are involved in the execution of a confirmed MMS service.

Step	Description
1	The client requests a service from the local MMS Provider.
2	The MMS Provider creates the appropriate MMS PDU and requests from the underlying layer (Presentation Layer, Layer 6) the transmission of the MMS PDU (not shown in the Figure).
3	The MMS Provider of the remote station (on which the server is running) receives the PDU and creates the MMS service indication which it delivers to the server.
4	The server executes the requested MMS service (e.g. starts the execution of a NC-Program) and sends the result of the service request to its local MMS Provider using the MMS-Service.Response primitive.
5	The MMS Provider of the server creates the MMS-Response PDU and requests from the underlying Presentation Layer the transmission of this PDU to the client station.
6	The Provider of the MMS client receives the MMS-Response PDU and sends the result of the service request to its local client using the MMS-Service.Confirmation primitive.

Table 3.1: Logical Communication Sequence for Confirmed Services

3.3.2. Unconfirmed services

Unconfirmed services are invoked by servers only. This class of service enables a server to notify a client that a predefined event has occurred. With this kind of service it is possible to avoid time consuming operations such as periodically reading the value of variables through the network. With unconfirmed services, polling is delegated to the server that performs the polling locally. MMS introduces three unconfirmed services: UnsolicitedStatus, InformationReport and EventNotification. A powerful functionality is defined in MMS to specify the conditions in which event notifications are sent. These services however are not yet implemented by device manufacturers. The user is thus forced to rely on UnsolicitedStatus and InformationReport to notify significant events. Figure 3.8 illustrates the steps that compose an unconfirmed MMS service.

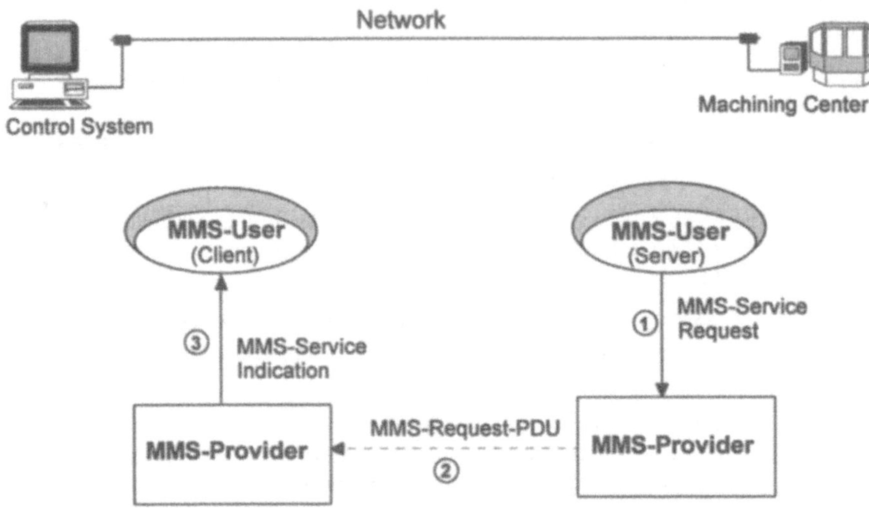

Figure 3.8: Steps of an Unconfirmed Service

3.4 The MMS-related standards

MMS is a standard application protocol that was jointly defined by the International Standardisation Organisation (ISO) and by the International Electrotechnical Commission (IEC). The standard is composed of two parts, the first defining the MMS service [ISO/IEC 9506-1], the second specifying the protocol [ISO/IEC 9506-2]. These two documents form what is sometimes called the **MMS Core** that is application-independent and can be applied to the definition of subsets of services and objects for specific devices. The documents in which this subset definition is made are called MMS **Companion Standards** and are being written or have been written for Robot Controllers, [ISO/IEC 9506-3], Numerical Controllers [ISO/IEC 9506-4], Programmable Logic Controllers [ISO/IEC 9506-5] and Process Control [ISO/IEC 9506-6]. The different parts of the MMS standard are shown in Figure 3.9.

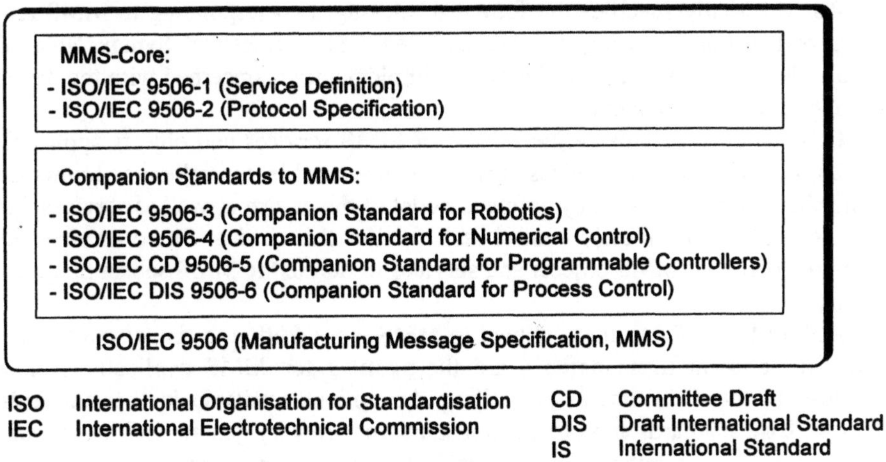

Figure 3.9: The different parts of the MMS standard

For users of MMS, Part 1 is the most important one. This part defines a set of object classes and the corresponding services that can be used to manipulate the objects of these classes. Figure 3.10 summarises the objects and services defined in Part 1 of the standard.

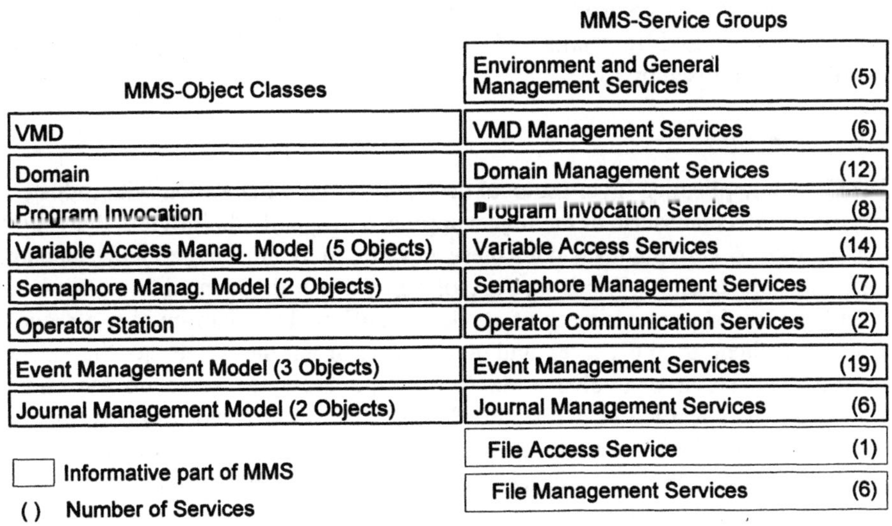

Figure 3.10: MMS object classes and services

Part 2 actually specifies the format of messages corresponding to MMS requests and responses, using a notation called ASN.1 (Abstract Syntax Notation One). This part is primarily of interest to implementors concerned with the development of MMS providers.

Companion standards define subsets of MMS services and objects applicable to families of industrial devices such as programmable controllers or numerical controllers of machines. They define a model of the device as well as services that can be applied to remotely control and monitor this device. These services are then mapped onto MMS objects and services to define the MMS subset that must be used.

A third kind of document related to MMS - in addition to the base standards and the companion standards - are the **profiles** for MMS applications. The purpose of these documents is to limit the options available in the standards. Indeed, base standards and companion standards are very general and leave many options in the definition of the services, for example optional parameters. Implementation by two different vendors could lead to products that are not able to communicate with one another. Profiles define which services and parameters must be present and which are optional. International Standard Profiles are being defined for MMS General Applications, NC and RC applications by regional organisations like EWOS (European Workshop on Open Systems), NIST OIW (OSI Implementors Workshop) and AOW (Asian OSI Workshop). International harmonisation of these specifications is one of the goals that these organisations are pursuing so that controllers produced by vendors of one region can be used with computers sold by vendors of another region. The ultimate goal is to have profiles standardised by ISO.

3.5 Summary

MMS is an application layer protocol designed for communication with industrial devices. MMS makes use of the object approach to specify the services that a user can invoke to communicate with these devices. It is an international standard, defining objects and services that are of general use for industrial applications. MMS distinguishes two classes of services: confirmed services involve a request from the client to the server and a response from the server to the client; unconfirmed services involve a single message, from the server to the client. MMS itself is specified in two standards: ISO/IEC 9506-1 and ISO/IEC 9506-2. Companion standards to MMS define how MMS can be used to communicate with special families of industrial devices. Companion standards have been defined for robot controllers, numerical controllers of machines and are being defined for programmable logic controllers as well as process controllers. A number of standards bodies are standardising profiles for MMS, following the same classification as companion standards. Harmonisation of these profiles by the different regional standards bodies ensures that industrial devices supplied by different manufacturers are able to interoperate, whatever their country of origin.

4. MMS Core Object Classes

In practice, it is neither necessary nor realistic to make use of all objects and services defined in MMS. Users as well as vendors have to negotiate which services and object classes should be used for an MMS implementation. As in most of the cases users have to integrate different device controllers, for example NCs, RCs, PLCs from different vendors, a good knowledge of the most important MMS classes and services is necessary. The following sections describe the VMD, Domain, Program Invocation and Named Variable object classes. These shall be referred to as the *Core* MMS object classes. Great importance is attached to the practical use of MMS which is reflected in numerous examples. These examples should help the reader understand the basis of MMS.

4.1 Virtual manufacturing device (VMD)

The VMD (Virtual Manufacturing Device) is an essential object in MMS. A VMD is an abstraction of a real manufacturing device (RMD). It is an MMS object and thus is described in terms of attributes. Abstraction means that the VMD hides implementation details, hiding differences between devices of a same family or of devices of two different suppliers. The VMD is the standard view that a client has of a server.

The essential benefit of this abstraction is that an RMD R2 can replace an RMD R1 as long as its VMD exhibits the same attributes, services and behaviour as the VMD of R1. This is very important for the user who can replace a device by another - maybe of a different supplier - or can reuse an application program written for this VMD but with different RMDs.

The VMD contains MMS objects that represent physical and logical resources of the RMD. It is a name space in which no two objects of the same class can have the same name.

The state and behaviour of a VMD is determined by the values of all attributes of the VMD, including all the attributes of contained and subordinated objects (Domain, Program Invocation, etc.). A VMD is static in nature as it cannot be created or removed from the system by MMS services.

Generally, the capabilities of the whole real device may be grouped into autonomous units and mapped onto more than one VMD. For example, a flexible machining centre may contain a tool magazine, a pallet store and even a tool measurement unit. It is possible, and in some cases it may be sensible, to map the resources and functionality of each of those units to different VMDs to reflect their autonomy. It is obvious, that there is a need for synchronisation of these units and therefore of the associated VMDs. As there is no definition and common understanding of how to synchronise different VMDs in MMS, in practice existing implementations map the capabilities of the real device onto one VMD.

The following services are defined on the VMD object:

- Status, to obtain the physical and logical status of the device
- UnsolicitedStatus, for the server to notify clients of changes in the physical and logical status of a device.
- Identify, for a client to read the name of the manufacturer, the version of the VMD, etc.
- GetNameList, to obtain a list of objects residing in the VMD.

4.1.1 Attributes of the VMD

Figure 4.1 gives the MMS definition of a VMD. The attributes of this object can be classified in three groups: descriptive attributes (Vendor and model name, revision), state attributes (logical and physical status) and attributes listing subordinate objects (list of capabilities, domains, program invocations, etc.).

Object:	VMD	
	Key Attribute:	Executive Function
	Attribute:	Vendor Name
	Attribute:	Model Name
	Attribute:	Revision
	Attribute:	List Of Abstract Syntaxes Supported
	Attribute:	Logical Status (STATE-CHANGES-ALLOWED, NO-STATE-CHANGES-ALLOWED, LIMITED-SERVICES-PERMITTED, SUPPORT-SERVICES-ALLOWED)
	Attribute:	Physical Status (OPERATIONAL, PARTIALLY-OPERATIONAL, INOPERABLE, NEEDS-COMMISSIONING)
	Attribute:	List Of Capabilities
	Attribute:	List Of Program Invocations (PIs)
	Attribute:	List Of Domains
	Attribute:	List Of Transaction Objects
	Attribute:	List Of Upload State Machines (ULSM)
	Attribute:	Lists Of Other VMD-specific Objects
	Attribute:	Additional Detail

Figure 4.1: Attributes of the VMD object

Vendor name, model name, revision

The content of the attributes Vendor Name, Model Name and Revision are specified by the vendor and contain identification information. For example, in a Siemens NC of the SINUMERIK family: Vendor Name = "Siemens-AG"; Model Name = "SINUMERIK 880", Revision = "Version 1.3". This information may be used by the client application to identify the remote device controller. The service Identify is used by the client to request this information from the server.

Example

The way to use the Identify service is shown in Figure 4.2. Identify is a confirmed service, which provides the identification information of the MMS server.

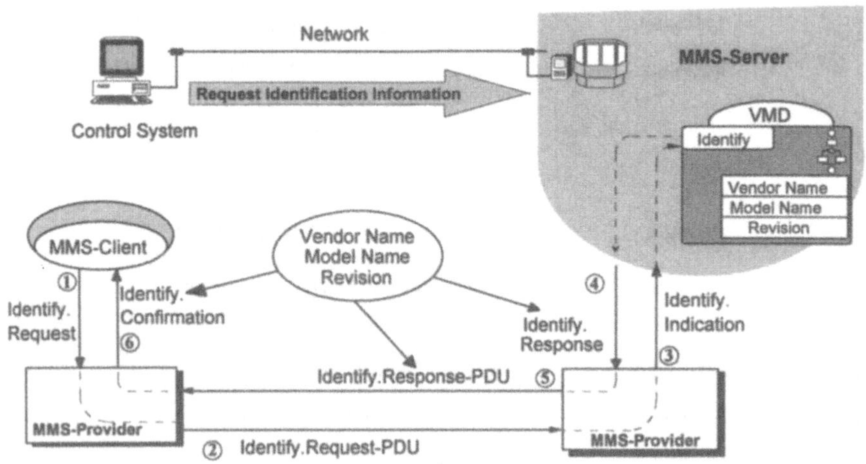

Figure 4.2: Request of Identification Information

Logical status

The attribute Logical Status reflects the actual level of functionality available through MMS. There are four separate levels as listed in Table 4.1.

Level	Value of Logical Status	Performed Services
1	STATE-CHANGES-ALLOWED	All MMS services which are supported by the MMS server implementation.
2	NO-STATE-CHANGES-ALLOWED	In general, all MMS services which do not cause any state changes or modifications of the actual value of any attribute of implemented MMS objects.
3	LIMITED-SERVICES-PERMITTED	Abort, Conclude, Status and Identify
4	SUPPORT-SERVICES-ALLOWED	All MMS services supported by the server implementation except the Program Invocation Management services Start, Stop, Reset, Resume and Kill.

Table 4.1: Possible values of the Logical Status

Physical status

The actual operational state of the hardware of the whole real device is characterised by the value of the attribute Physical Status.

Value	Operational state of the real device
OPERATIONAL	No deficiencies are detected. The real device is able to perform its intended task.
PARTIALLY-OPERATIONAL	At least one hardware malfunction is detected and therefore limitations of the functions supported by the device may exist.
INOPERABLE	The real device is unable to do any useful task.
NEEDS-COMMISSIONING	Local commissioning process is necessary.

Table 4.2: Possible values of the Physical Status

4.1.2 A simple example

Figure 4.3 illustrates a scenario in which a client requests the status of a manufacturing device. Calling the MMS Status service results in a message being sent from the client to the server. The latter interprets the service and executes it. The result, composed of the physical and logical statuses is passed to the service provider of the server which constructs a response and sends it back to the client.

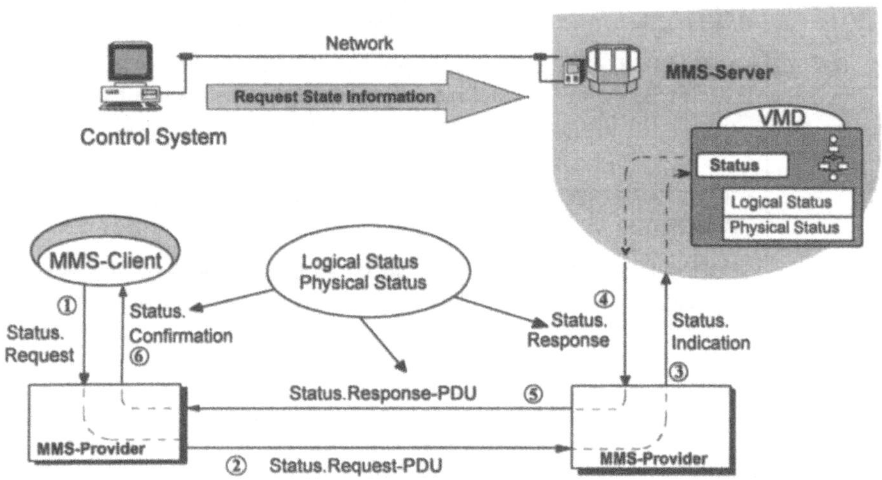

Figure 4.3: Request of the actual state of the manufacturing device

4.1.3 The UnsolicitedStatus service

The following scenario is illustrated in Figure 4.4. A local operator of a machine tool wants to modify an NC program. In order to avoid changes via the network it is necessary to switch the operation mode of the device controller from Remote to Local. This state of the machine tool is reflected by the VMD attributes Logical Status and Physical Status with the actual values NO-STATE-CHANGES-ALLOWED and OPERATIONAL respectively. As the client does not know when the state change occurs, it has to periodically poll the device to detect a change of state. Therefore the client sends a Status request to the server in a cyclical way. To avoid this time-consuming operation, MMS defines the UnsolicitedStatus service which allows the server to report the change of state to the client. It is important to note that the UnsolicitedStatus service is used by the server and not by the client. Furthermore this service is an unconfirmed service, which means that no response or confirmation is expected from the client.

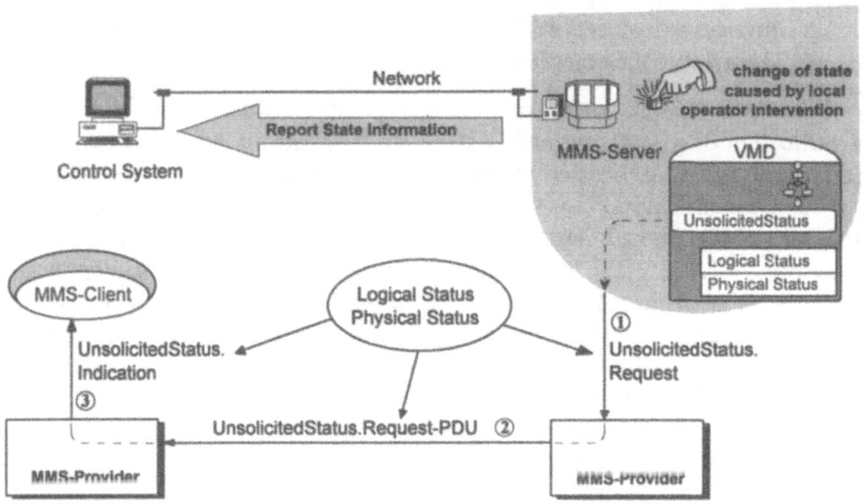

Figure 4.4: A device reporting a change of state

4.1.4 Other attributes

The attributes List Of Abstract Syntaxes Supported, List Of Capabilities and Additional Detail are not discussed here as they are not relevant for the practical use of MMS.

The reference attributes List Of Program Invocations, List Of Domains and List Of Other VMD-specific Objects are introduced in detail after the discussion of the objects Domain, Program Invocation and Named Variable.

4.1.5 Naming in MMS

MMS distinguishes two categories of objects: named and unnamed objects. The majority of object classes are named, which means that objects of these classes are accessed by name and not by address. The name of an object is composed of two pieces of information:

1. The **scope** of the object. MMS introduces three scopes: VMD-specific, Domain-specific, Association-specific. A scope is essentially a name space, in which no two objects of the same class can have the same name.
2. An identifier of the object in this scope.

The VMD object is the only unnamed object considered in this book. Identification of a VMD is made not by an MMS name but by a **Presentation Address** or by an **Application Reference**. As a Presentation Address may change due to a reconfiguration or a relocation of a server, it is recommended to designate VMDs through Application References. This notion is defined precisely in section 7.6.4 of [ISO/IEC 9506-2]. In a simplified way, an application reference is a series of names and identifiers that allow to unambiguously reference a part of the VMD called **Application Entity**. Application references are used in a few MMS services that involve third parties. An example will be given later for the transfer of a domain.

4.1.6 Associations and the MMS environment

Prior to any confirmed or unconfirmed service invocation, an association must be established between the client and the server. An association can be seen as an application level connection that binds a client to a server. Communication on an association is bidirectional, in the sense that both the client and the server can invoke MMS services. Establishment of an MMS association results from invoking the Initiate service. Termination of an association can be orderly or abrupt. In an orderly termination, one of the communicating partners invokes the Conclude service. The association is considered terminated when the requesting MMS user receives a positive response from his peer. In the case of a negative response, the association is still considered established. Abrupt termination of an association does not require that the peer agrees to terminate. The service invoked in this case is the Abort service.

Concluding or aborting an association has an impact on the VMD of the server. All objects having an association-specific scope are removed from the VMD.

When an association is established between two MMS users, a negotiation takes place to determine the value of a number of parameters. These are the maximum number of pending services when calling or when called by the peer, the maximum nesting level of data structures, the version number, the parameter building blocks. In addition to this negotiation, both MMS users indicate to their peer the services they support.

The MMS environment is considered established in the requesting user when a positive response is received for the Initiate service while it is considered established in the responding user when the positive response is sent.

4.2 Domains

As mentioned in the description of the VMD, a real device may contain a set of capabilities which fit together. In order to reflect this unity MMS defines the Domain object to represent a subset of the resources of a real device, e.g. the tool magazine and the available tools of a flexible machining centre or the robot arm of a robot system. In general an object of the class Domain may contain every kind of data, which can be uploaded in order to store the image in an external system, from where the uploaded image may be downloaded to the device controller when it is needed. The content of the domain may either be visible or invisible to the client.

In practical manufacturing applications a domain contains either objects of other classes, e.g. Named Variables or may be used as a container which contains instructions for the controlling program of a numerically controlled manufacturing device, for example NC part programs. These instructions (programs) can be transferred to and from the programmable manufacturing device by the use of MMS services defined for the class Domain.

There are three distinct ways to create a domain:

1. A domain may come into existence during a download procedure as shown in the example "Domain Download Procedure".
2. Domains may be predefined in the system, e.g. created during the system setup procedure.
3. A domain may also be created by local means, e.g. as part of the execution of a Program Invocation.

The following sections give a definition of the Domain object class. A number of examples show the practical use of domain objects and services.

4.2.1 Attributes of the domain object

A Domain can be seen in two ways: as a container for code and/or data used by Program Invocations and as a collection of other MMS objects, such as variables. In the first viewpoint, download and upload services are important. In the second viewpoint, the domain-specific scope of objects is the main aspect. Figure 4.5 gives the MMS definition of a Domain. The following paragraphs describe the most important attributes of this object.

Object:	Domain		
	Key Attribute:	Domain Name	
	Attribute:	List Of Capabilities	
	Attribute:	State (LOADING, COMPLETE, INCOMPLETE, READY, IN-USE, D1...D8)	
	Constraint:	State = (LOADING, COMPLETE, INCOMPLETE, D1, D2, D3, D9)	
		Attribute:	Assigned Application Association
	Constraint:	State = (READY, IN-USE, D4, D5, D6, D7, D8)	
		Attribute:	Upload In Progress
	Attribute:	MMS Deletable (TRUE, FALSE)	
	Attribute:	Sharable (TRUE, FALSE)	
	Attribute:	Domain Content	
		Attribute:	List Of Subordinate Objects
	Attribute:	List Of Program Invocation (PI) References	
	Attribute:	Additional Detail	

Figure 4.5: Attributes of the Domain object

Domain name

This key attribute contains the name of the domain object. Each domain included in a VMD is uniquely identified by its name. The physical address of the domain is not visible. Note also that domains cannot be included in domains, which means that domains have either VMD-specific or association-specific scope.

State

One of the most important attributes of the domain is the State attribute. The value of this attribute specifies the actual state of the domain as defined by the Domain State Diagram. The most important states are LOADING, READY and IN-USE. The intermediate states COMPLETE, INCOMPLETE and D1 to D9 are not significant for the description of the Domain State Diagram, as shown in Figure 4.6 and therefore are not considered in Table 4.3. This table contains a short description of the three significant states mentioned above.

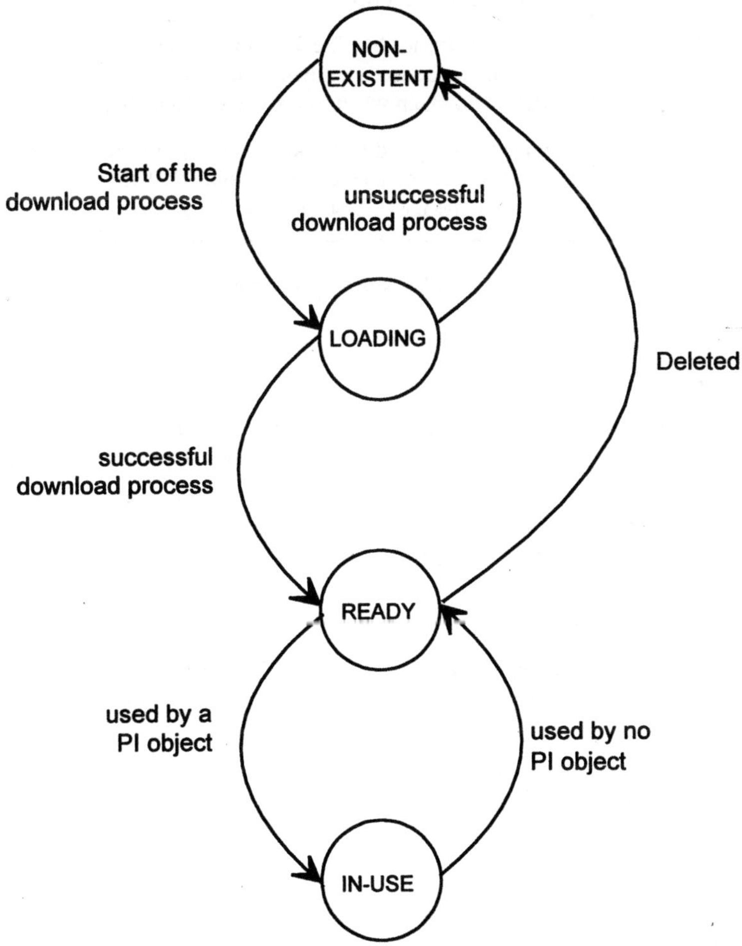

Figure 4.6: Domain State Diagram (reduced)

State	Short description
LOADING	In this state, the domain is currently involved in a download procedure. This state is only significant for a domain which is explicitly created as part of a download procedure. In this state, the integrity of the domain content is not guaranteed. If the download procedure is successfully finished the domain enters the READY state. Otherwise the domain is removed from the VMD.
READY	This state indicates that the domain is accessible. There is no distinction between a downloaded domain and a predefined domain which is created by local means.
IN-USE	If the domain is used by (linked to) one or more Program Invocations, the domain enters the state IN-USE. A domain which is in the IN-USE state cannot be deleted.

Table 4.3: Main states of a domain

Upload in progress

This conditional attribute contains the number of active upload procedures. If there is no upload procedure active for this domain, this attribute is not visible. This attribute is necessary as the Domain State Diagram does not consider active domain upload procedures and a domain can not be deleted, if an upload procedure is underway for this domain.

MMS deletable

A domain may be deleted by a client if and only if the value of this attribute is TRUE. If the value is FALSE, the client is not allowed to delete this domain. In practice, all predefined domains and domains which are created by local means are not deletable using an MMS service. Therefore the value of the attribute MMS Deletable of such domains should be FALSE.

Sharable

As mentioned in the description of the Domain State Diagram, a domain may be used by one or more PIs (State = IN-USE). If the value of the attribute Sharable is FALSE the domain may be used by a single PI. If the value of the attribute Sharable is TRUE, the object may be used by multiple PIs. The concurrent use of a sharable domain by more than one PI must be synchronised by local means.

List of subordinate objects

This attribute references all objects which are part of the domain and which are visible to the client. These subordinate objects come into existence with the creation of the domain and are deleted when the domain is deleted.

In practice, domains with subordinate objects are pre-existent and contain structured information which is accessible by the MMS client. As they are pre-existent they are expected to be non-deletable (MMS Deletable = FALSE). Furthermore, such domains will not be uploaded or downloaded by the MMS client. They are used as containers for the subordinated and externally visible objects, for example Named Variables.

List of program invocation references

All Program Invocation objects which use the domain are included in this list of references. The values of the attributes State, Sharable and List of Program Invocations References are interdependent. This interdependence is shown in Table 4.4.

List Of Program Invocation References	State	Sharable
empty	READY	TRUE / FALSE
only one entry	IN-USE	FALSE
one or more entries	IN-USE	TRUE

Table 4.4: Interdependence of three domain attributes

The transfer of untyped data, for example programs or process images, to and from a manufacturing device is one of the most important procedures in practice. Therefore the handling of domains is essential for the user. The following examples demonstrate the transfer of such data to and from a numerically controlled manufacturing device using MMS services.

There are two different ways to initiate the transfer of data. Either the client initiates the download or the upload procedure, or the server requests from the client to start the download or upload procedure. Both cases are covered by MMS and illustrated by the following examples.

4.2.2 Domain download initiated by the client

This example shows the use of MMS to download an NC program from a control station to a device controller (NC). In this example, the download procedure is initiated by the client. The NC program is stored in a file of the control station.

Table 4.5 contains a description of the essential procedures followed by the client and server and the MMS services used. As this example illustrates a successful download procedure, no errors and therefore no negative service responses are considered.

Step	MMS Client	MMS Server
1	Initiates the download procedure using the InitiateDownloadSequence service.	
2		Receives the indication of the service and creates a domain. The attributes Domain Name and Sharable get the value received from the client. The value of the attribute MMS Deletable is set to TRUE. The domain enters the state LOADING.
3		The server sends a positive response.
4	After receipt of the confirmation the client waits for the request to transfer the first segment.	
5		The server requests the transfer of the data using the MMS DownloadSegment service. Thus the server controls the data flow. A single segment can be transferred at a time.
6, 7	The client may open and lock the local file which contains the NC Program. It reads the first segment[1] which is sent to the server. Additionally the Client informs the Server if there are more segments.	
8		The server may check the content of the first segment and stores the received data in a local data store (e.g. local file system, local RAM).

[1] The number of bytes which can be transferred at once are limited by the maximal size of a MMS PDU which is negotiated between the MMS Client and Server. Therefore a segmentation of the whole file content may be necessary.

9, 10		The procedure is continued until the client sends the last segment.
11, 12	When the client reads the last segment of the file, it sends the last data and informs the server that there are no further data available. The client waits for a message indicating completion of the download.	
13		The server stores the data of the last segment. As there are no more data to transfer, it terminates the download procedure[2].
14		The server terminates the download procedure and informs the client about the result of the download procedure.
15	After receipt of the TerminateDownloadSequence, the client decides whether the domain should remain on the server (Response(+)) or should be deleted (Response(-)).	
16		If the server receives a positive response of the termination, it changes the state of the domain to READY. Otherwise any loaded data is discarded and the domain is deleted.

Table 4.5: Download of a domain

---

2 It should be mentioned that the check of the content of the transferred data is a local matter. In one implementation there may be a check of each received segment in an other implementation there may be only a check of the whole content of the Domain object after the receipt of the last segment. In some cases there may be no check. This should be kept in mind.

The steps in the first column of the table above correspond to the circled numbers shown in Figure 4.7.

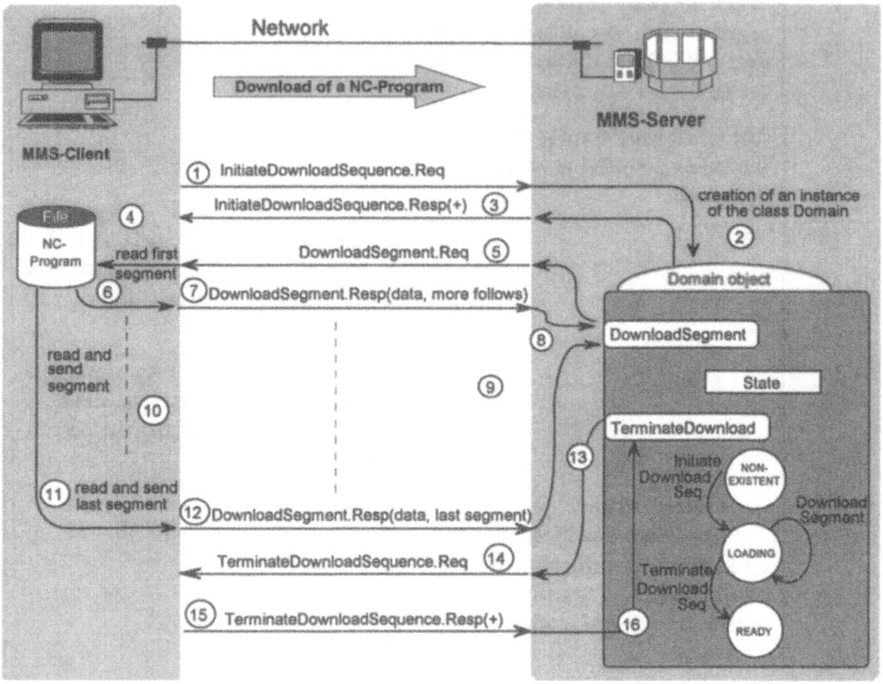

Figure 4.7: Download of an NC program

4.2.3 Domain download initiated by the server

The following example, in Figure 4.8 describes how the server asks the client to start the download of an NC program. The starting of a download may be caused by local actions, for example, a local operator requiring an NC program.

After receipt of the request from the server to initiate the download, the client starts the procedure as described in the above mentioned example. The server requests the starting of the download using the MMS RequestDomainDownload service.

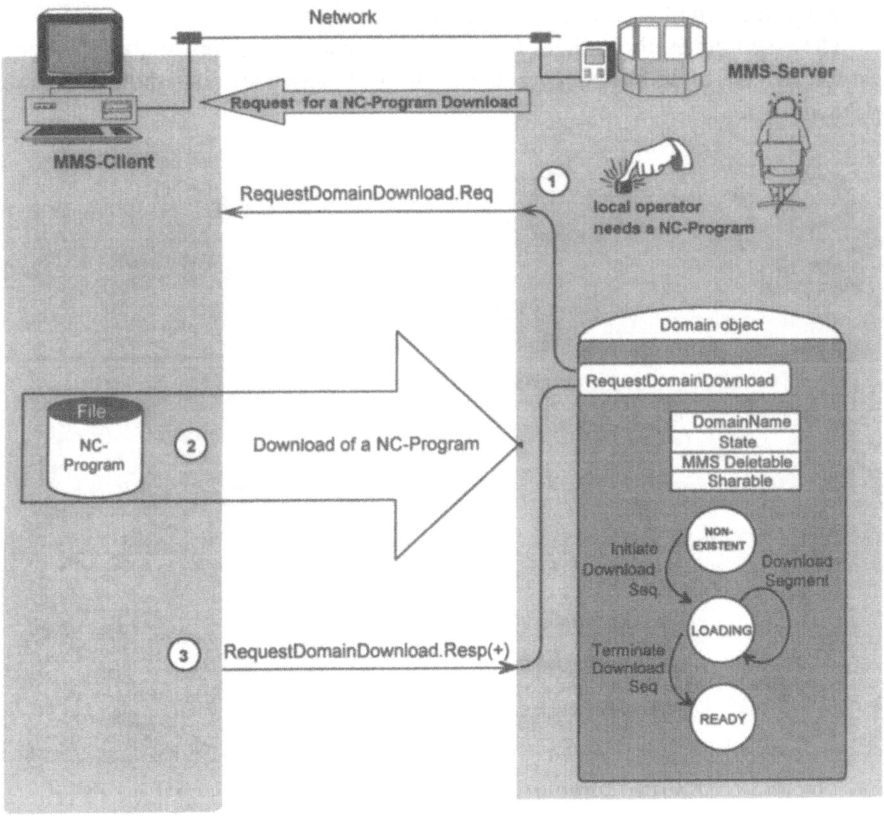

Figure 4.8: Download of an NC program requested by the server

Note that the response to the RequestDomainDownload service is sent by the client at the end of the download, and at this moment only.

4.2.4 Domain upload

In general, a device controller has an operator station - terminal and display - to enable the operator to locally control the device and the execution of programs. In some applications it is necessary to modify a program or data which has been downloaded. The operator may also create new programs using local facilities, for example a local NC programming system.

This requires the possibility to upload changed or created programs or data from the device to a remote station. As already mentioned, programs and untyped data are mapped onto MMS domains. MMS defines a set of services for the upload procedure of programs and data stored in domains.

The client generally initiates the upload procedure. The server may also request the starting of an upload from the client. Both cases are illustrated in Figures 4.9 and 4.10.

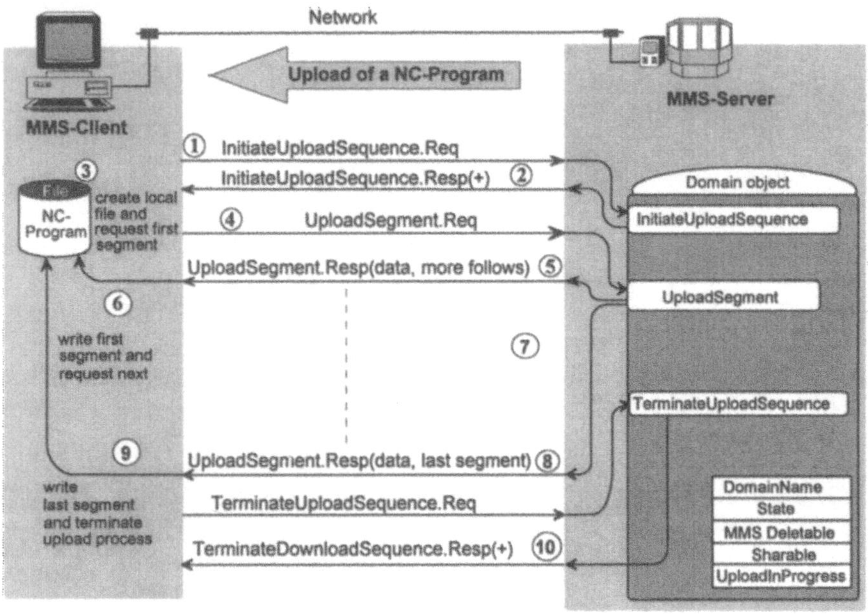

Figure 4.9: Upload of an NC program initiated by a client

Uploading of an NC program can also be initiated from a server as shown in Figure 4.10.

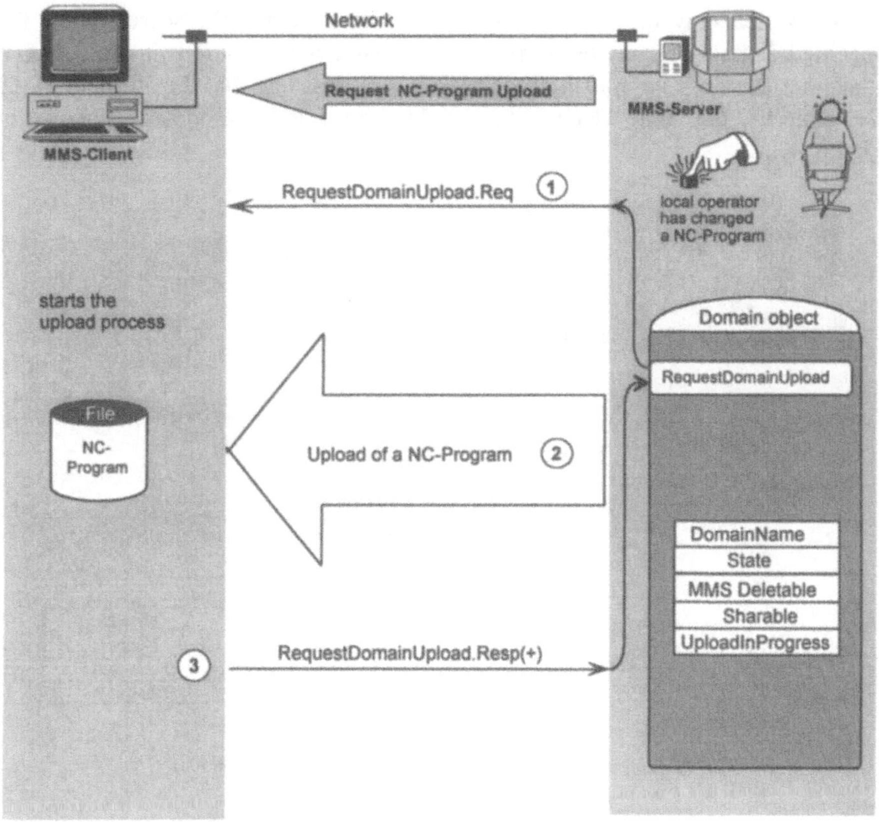

Figure 4.10: Upload of an NC program requested by a server

4.2.5 Deletion of a domain

A domain which came into existence as part of a download procedure may be deleted by the client, if it is not needed anymore. MMS defines the DeleteDomain service which may be used by the client to request that the server delete the specified domain. The server verifies that the specified domain is in the READY state, that there are no upload procedures in progress and that the attribute MMS Deletable has the value TRUE. The sequence for deleting a domain is shown in Figure 4.11.

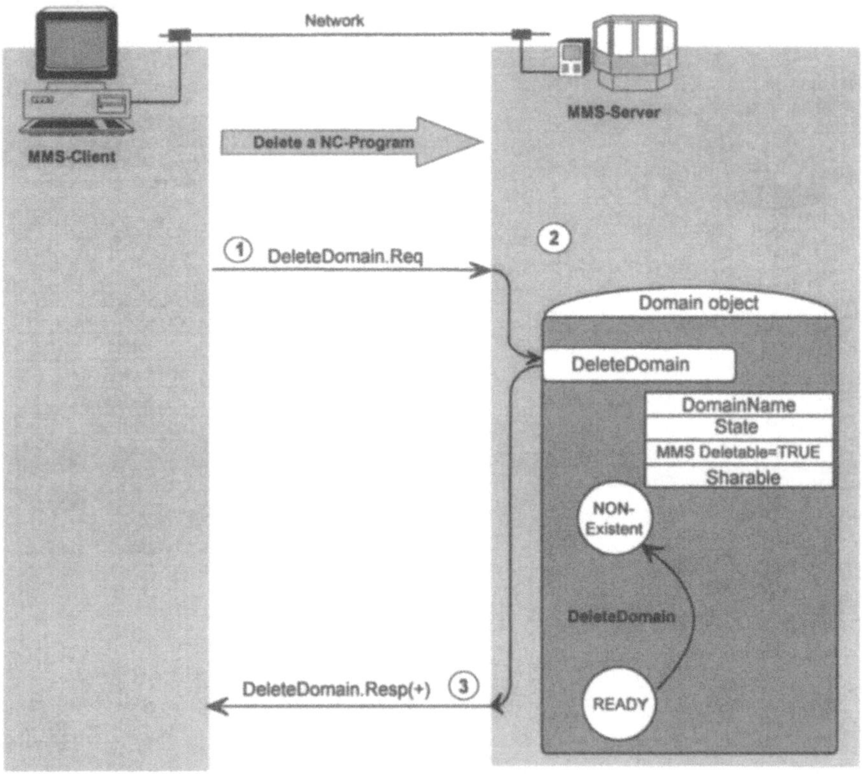

Figure 4.11: Deletion of a domain

4.2.6 Getting the attributes of a domain

In general, the result of the execution of an MMS service depends on the actual value of one or more attributes of an object.

For example, the use of the DeleteDomain service is only meaningful if and only if the attribute State of the domain has the value READY. If the client wants to delete a domain, it should check the actual values of the attributes of the domain. In order to avoid the request of an MMS service which results in a negative confirmation, MMS defines services to get the actual values of the attributes of an object. The client may use the GetDomainAttributes service in order to obtain the values of the attributes of a domain. Figure 4.12 illustrates the use of the GetDomainAttributes service. For performance reasons it may be sensible to send a DeleteDomain service request without checking the current state with the GetDomainObjectAttributes service. In such a case the client must be able to handle a negative confirmation, if the domain is not in the READY state or if it is involved in an upload procedure.

Furthermore, this service may be used by the client to obtain the current picture of the domain and hence decide what to do next according to its current control plan.

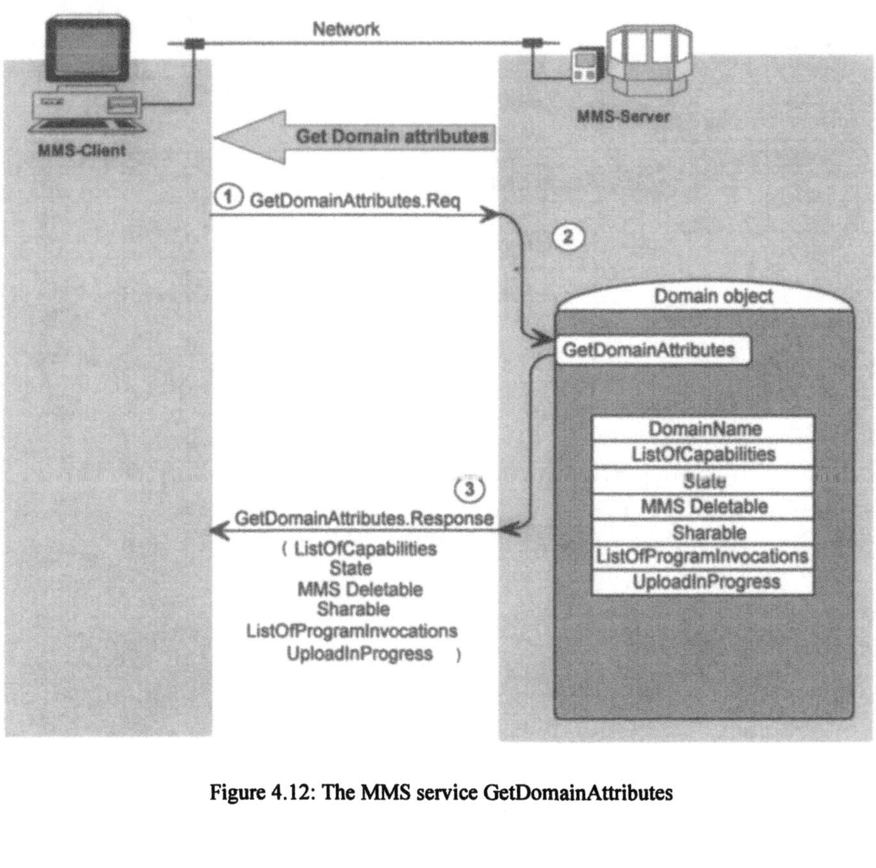

Figure 4.12: The MMS service GetDomainAttributes

4.2.7 Practical example

The following example shows a standard sequence of operations which may often be found in a computer aided manufacturing environment. An operator attending a machining centre has to verify a new NC program which may be created by a remote NC program supply system. After the verification, the NC program may be marked as tested and therefore as available for the automated manufacturing system. He requests from the remote station, for example an NC program supply system, the NC program which must be verified. In order to avoid state changes caused by remote requests, the operator has to change the operation mode from remote to local mode. This is done by local means, e.g. by an operation-mode-switch of the local control panel. Before the operator may start the execution of the NC program, he has to check and if necessary to modify the program. At the end of the verification he requests the upload of the modified NC program in order to save the modifications. The operation mode is then changed from local to remote in order to permit the remote control of the machining centre.

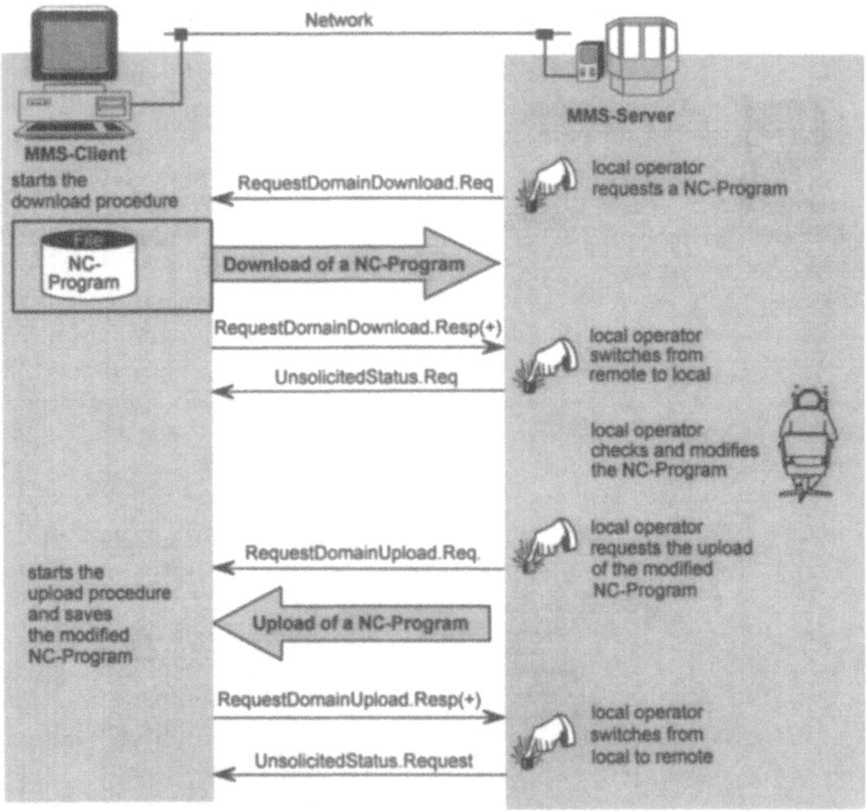

Figure 4.13: Download, verification and upload of an NC program.

The use of MMS for this scenario is shown in Figure 4.13. It should be mentioned that after this procedure the domain still exists on the server. This domain should be removed either by the client using the DeleteDomain service or by the server using local means, if it is no longer needed.

4.3 Program invocations

The term Program Invocation has already been introduced to the reader earlier in this book but now a more detailed description of this MMS object class is provided.

An object of the class Program Invocation may be seen as the description of the externally visible behaviour of a task or an execution thread in a multi-tasking environment. MMS defines this class to describe the externally visible life cycle of an NC program or other sets of statements executable by the device controller. In a numerical controller, an NC program represents a set of statements which are used by the controlling program of the controller as input for motion control of the axes or other components of the manufacturing device. The execution of an NC program is described by a set of states and transitions.

An object of the class Program Invocation may be dynamic in the sense that it comes into existence and may be removed from the system. A Program Invocation may come into existence either by MMS services or by local action, or it may be predefined. PIs are composed of one or more domains that contain the code and data of the program. A domain containing the code of a PI can be shared by several Program Invocations.

This model of program invocations is very general and is similar to programs that run on typical general purpose computer systems as well as programmable manufacturing controllers. For instance, most application programs for computers are actually domains that must be first loaded into memory. The program invocation that is executed consists not only of the domain containing the application program, but also the run time system domain. You usually cannot run the application program by itself, the run time system needs to be bound to the application program before execution can take place. This binding of sub-programs to create an executable program is what MMS calls a Program Invocation (PI).

4.3.1 Attributes of a program invocation

Figure 4.14 lists the attributes that characterise an MMS Program Invocation. All attributes except the State and the Execution Argument of the PI are initialised at creation time. The State is the only attribute that can change during execution of a PI while the Execution Argument can be modified only when starting or resuming the PI. The format of the latter parameter is not defined by MMS and is vendor dependent.

Object:	Program Invocation		
	Key Attribute:	Program Invocation Name	
	Attribute:	State (IDLE, STARTING, RUNNING, STOPPING, STOPPED, RESUMING, RESETTING, UNRUNNABLE)	
	Attribute:	List Of Domain References	
	Attribute	MMS Deletable	
	Attribute:	Reusable	
	Attribute:	Monitor (TRUE, FALSE)	
	Constraint	Monitor = TRUE	
		Attribute	Event Condition Reference
		Attribute	Event Action Reference
		Attribute	Event Enrolment Reference
	Attribute	Execution Argument	
	Attribute	Additional Detail	

Figure 4.14: Attributes of a Program Invocation

Program invocation name

Each instance of the Program Invocation class is unambiguously identified by its name. This key attribute contains the name of the Program Invocation object.

State

MMS defines a Program Invocation State Diagram, as shown in Figure4.15. The actual state of the PI is reflected by the value of the State attribute. The transient states STARTING, STOPPING, RESUMING and RESETTING are not considered in the figure. These states represent situations in which the VMD has received a Start, Stop, Resume or Reset request, has started executing them but has not completed their execution. Given the fact that there may be cases in which the state transitions take a non negligible time, it may be necessary to take into account these transient states as well. This book does not consider such cases.

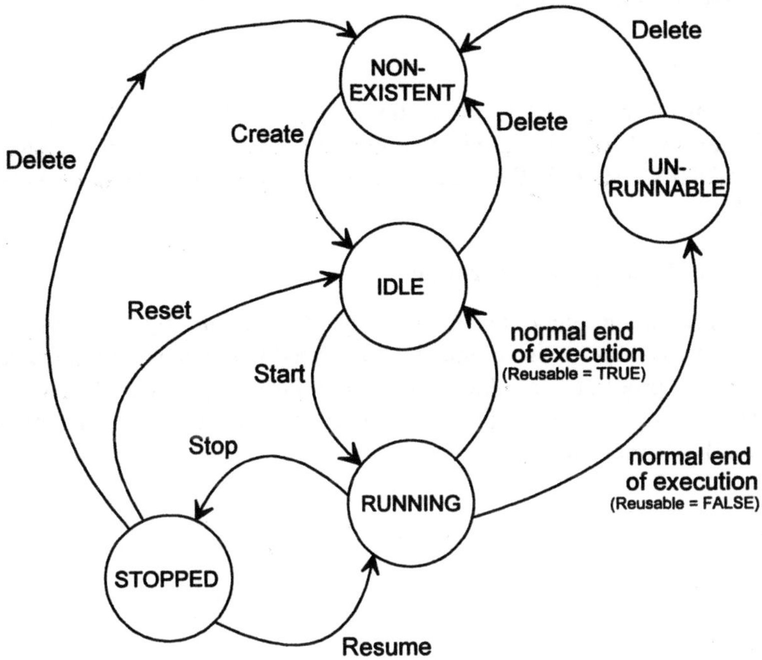

Figure 4.15: Program Invocation State Diagram (reduced)

List of domain references

This reference attribute is a list of references to the domains which compose this Program Invocation. It should be mentioned that this list must contain at least one entry. This means that a VMD which contains a Program Invocation contains at least one domain which is linked to it.

MMS deletable

The value of the attribute MMS Deletable indicates whether the Program Invocation can be deleted (MMS Deletable = TRUE) using the DeleteProgram-Invocation service or not (MMS Deletable = FALSE). The value of this attribute is TRUE if the Program Invocation was created by a client using the Create-ProgramInvocation service.

Reusable

This attribute indicates whether the Program Invocation can be placed in execution again or must be removed after the first execution. In many applications this attribute has the value TRUE.

Monitor

This attribute has the value TRUE if and only if the powerful Event Management facilities defined by MMS are used to inform the client whenever the object leaves the RUNNING state. If these facilities are not used, the attribute has the value FALSE.

4.3.2 Creation of a program invocation

A Program Invocation may be created by the client by use of the CreateProgramInvocation service or may be predefined in the system (e.g. created during the set-up procedure of the device controller). Figure 4.16 shows an example of the use of the remote creation of a Program Invocation by use of the CreateProgramInvocation service.

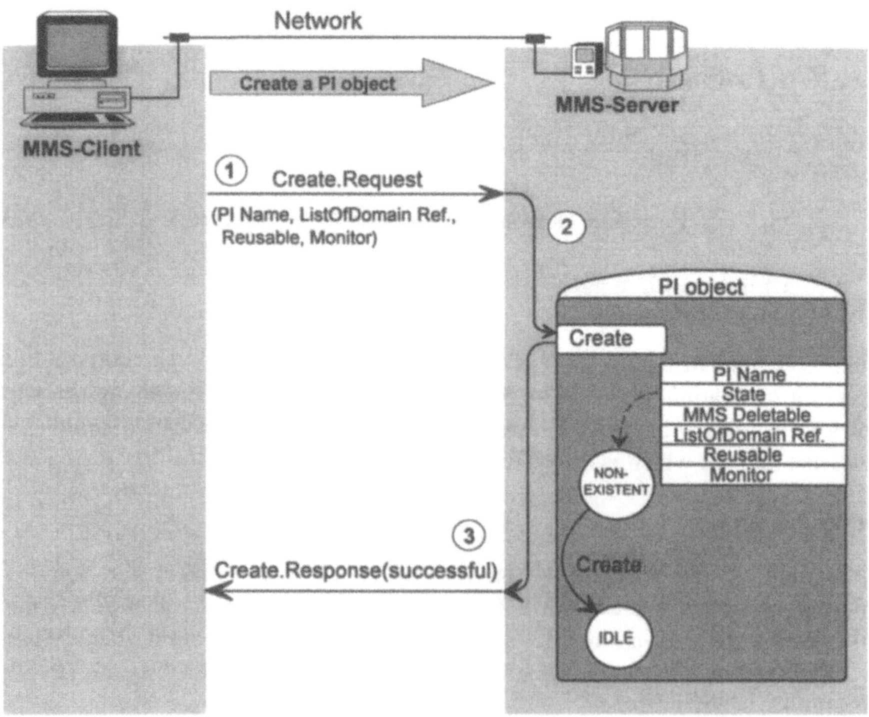

Figure 4.16: Creation of a Program Invocation

The server must verify that no Program Invocation already exists with the same name and that all domains listed in the parameter List Of Domain Names are available and that they are in the READY or IN-USE state. If a domain is already in the state IN-USE its attribute Sharable must be equal to TRUE. After a successful verification, the server creates the Program Invocation with its State attribute equal to IDLE and sends a positive response to the client. At this moment, a Program Invocation exists which may be started by the client.

4.3.3 Starting a PI

The client may remotely control the execution of an NC program via the network by using the Program Invocation services Start, Stop, Resume, Reset and Kill. The Start service is the most important one and its use is illustrated in Figure 4.17. For the Start service, there is no distinction between a local predefined Program Invocation and a Program Invocation created by the client. The Start service initiates the execution of a PI, for example an NC program.

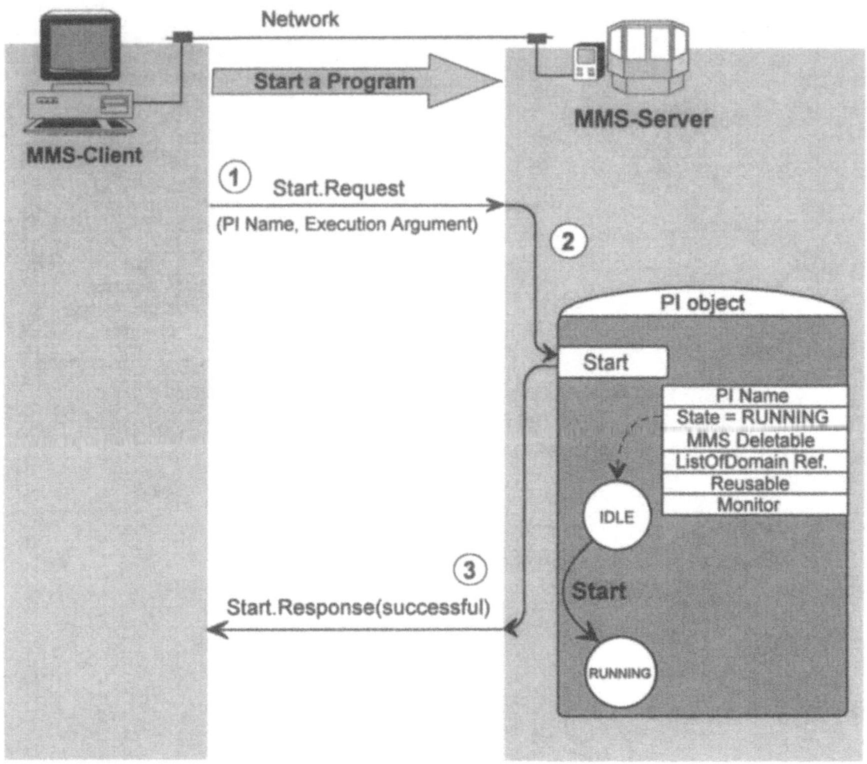

Figure 4.17: Starting an NC program by use of the Start service.

4.3.4 Obtaining the state of a PI

In some applications, the client needs information about an NC program, for example the current state of the program. The state of an NC program is reflected by the attribute State of the associated Program Invocation object. The current state of this object is accessible through the network by using the GetProgramInvocationAttributes service.

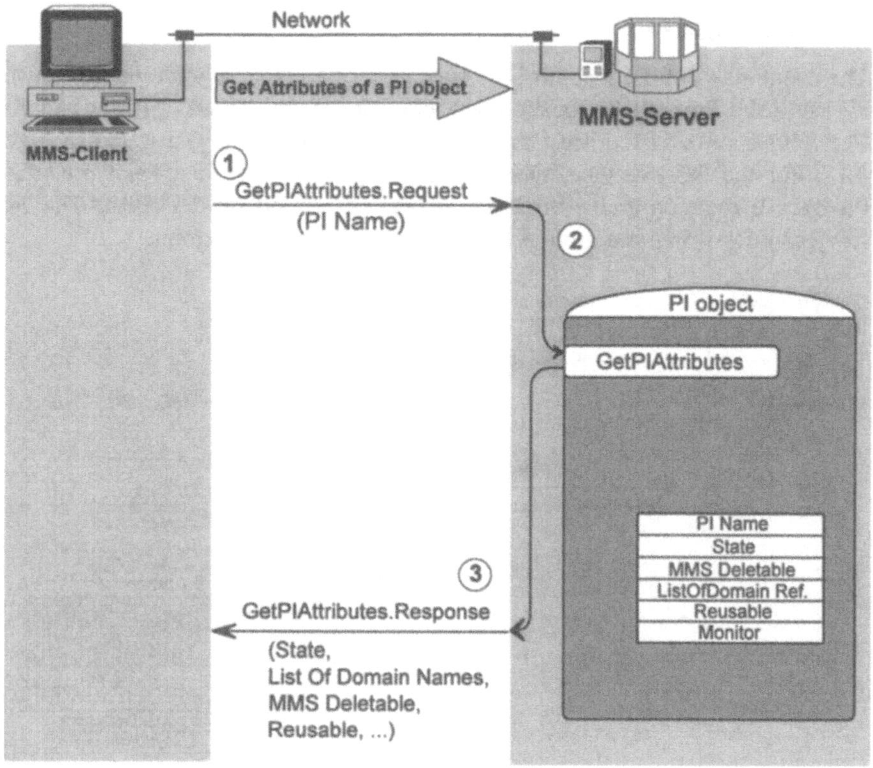

Figure 4.18: Get the state of an NC program

4.3.5 Deletion of a PI

As an NC program uses resources of the device controller, it is necessary to delete the program when this is no longer needed. The executable statements of the program may be stored in one or more domains which are associated to the Program Invocation. In order to free all resources used by the NC program, the client deletes the Program Invocation and if no longer needed, all domains referenced by the Program Invocation. As illustrated in Figure 4.19, the Program Invocation can be either in the IDLE, STOPPED or UNRUNNABLE states.

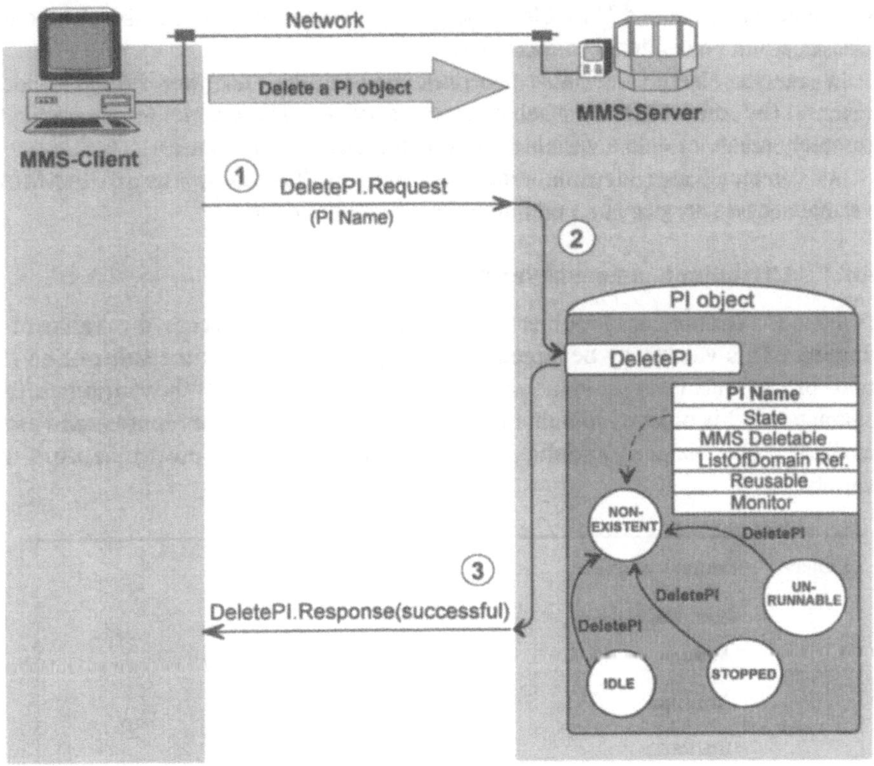

Figure 4.19: Deletion of a Program Invocation.

4.4 MMS variables

Most data of a device controller, that are accessible via the network, are represented by real variables, in the sense of concrete variables. A typical example in a manufacturing environment are the tool data that give the precise dimensions of a tool used by a machine. Real variables are mapped to MMS variables which may be seen as an abstraction to describe the external view of real variables. In order to access real variables, MMS defines the object classes Unnamed Variable, Named Variable, Scattered Access, Named Variable List and Named Type. Fourteen services are defined to handle these objects. In practice, the object class Named Variable is the most important one for the user. The other classes are not considered in this book.

In practice, Named Variables are pre-defined or their creation is done by local means. The other MMS variable access services that enable for example to remotely create or delete variables are not necessary in most cases.

In Chapter 6, we give numerous examples to illustrate the use of the MMS variable access services in a real manufacturing application.

4.4.1 Attributes of a named variable

A Named Variable is an object that describes how to access the value of a variable. This value may be stored in memory, may represent the value of an I/O port or the result of a value generator. Figure 4.20 lists the attributes that characterise this object. Note that the value of the variable is not represented as an attribute. The Named Variable is a descriptor indicating how to access the variable.

Object:	Named Variable		
	Key Attribute:	Variable Name	
	Attribute	MMS Deletable (TRUE, FALSE)	
	Attribute:	Type Description	
	Attribute:	Access Method (PUBLIC, ...)	
	Constraint	Access Method = PUBLIC	
		Attribute:	Address

Figure 4.20: Attributes of a Named Variable

Variable name

This attribute uniquely identifies the Named Variable object and may be defined with VMD-specific, Domain-specific or Application Association-specific scope. VMD-specific means that the Named Variable object is visible at the level of the VMD. If the object is part of a domain, its name is Domain-specific and contains the name of the domain as well as the name of the Named Variable object.

MMS deletable

This attribute indicates whether or not the object may be deleted by use of the MMS DeleteVariableAccess service. In practice, only Named Variable objects created by the MMS-client by use of the MMS DefineNamedVariable service may be deleted. The majority of Named Variable objects which are used in MMS applications are predefined and not deletable by MMS services.

Type description

This attribute contains a description of the data structure of the local variable, which is represented by the MMS Named Variable object. This attribute may describe a simple data type - boolean, integer, etc. - or a complex type. Section 4.4.2 describes the types that can be defined and used in MMS.

The other attributes are not relevant for the MMS user and therefore are not described in this book.

4.4.2 Data types

Named variables are typed in MMS. A type indicates the nature of the variable and the operations that can be performed on it. For example, integers support the traditional arithmetic operations while booleans support the logical operations: combining booleans with AND, OR, XOR operators. Table 4.6 summarises the basic types supported in MMS.

Complex types can be created by combining simple types with one another using two constructors: arrays and structures.

An **array** is an ordered collection of elements accessible by an index. These elements may be of a simple type or of a complex type - array and structure - but all elements must be of the same type.

A **structure** is an unordered collection of elements called fields, that are accessible by name. The elements may be of a simple type or of a complex type - array and structure - and may be of different types.

MMS Type	Description
Boolean	Two values - TRUE or FALSE
Bit string	String of binary data
Integer	Signed integer number
Unsigned	Positive integer number
Floating point	Real number in a simplified format
Real	Real number in a complex format - rarely supported
Octet string	String of octets
Visible string	String of printable characters
BCD	Binary representation of decimal numbers
Object Identifier	A special type for object designation
Binary time	Time represented as an integer number
Generalised time	Date and time of day

Table 4.6: Basic types for MMS variables

4.4.3 Alternate access

Arrays and structures are composed of multiple elements and can in some cases be very large. Sometimes, a client is interested in reading or writing only a part of these structures. For an array of seven elements for example, the client may be interested in elements 1, 3, 6 and 7. Two approaches can be adopted to read these elements:

1. The client reads the whole array and then extracts from the result of this operation the elements 1, 3, 6 and 7. The problem with this approach is that the whole array is transferred. In this particular case, this does not represent a problem as the array is short. However, for large arrays and structures with hundreds or thousands of elements, this may be time consuming.

2. The client specifies in the read request which elements must be read. In the example described above, the client specifies that he is interested in elements 1, 3, 6, 7 only. The server in this case returns these elements and these elements only.

The alternate access is a mechanism available in MMS that allows a client to specify which elements of an array or structure he wants to read or write. The elements may themselves be arrays and structures.

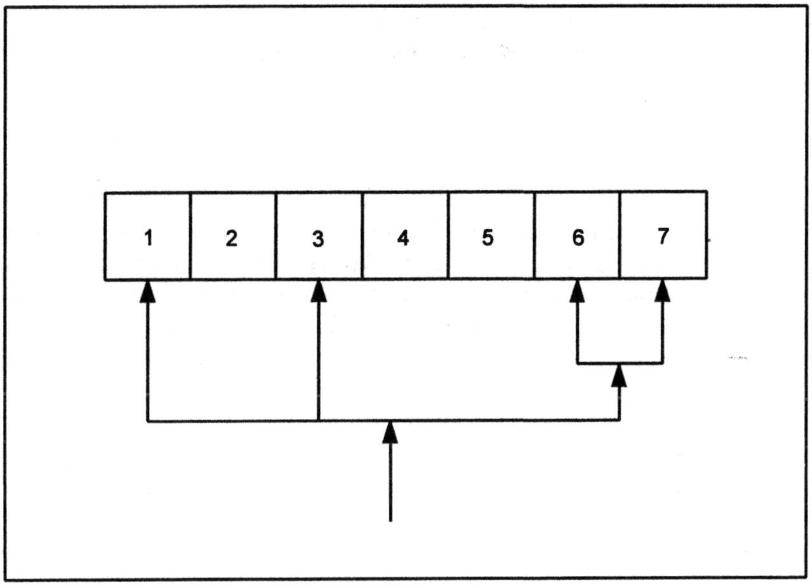

Figure 4.21: Example of alternate access for an array of seven elements

Figure 4.21 illustrates for the example described in this section the principle of the alternate access: a list of one or multiple descriptors, each specifying either a single element or a range - here the range is 6-7.

4.4.4 Getting the attributes of a variable

To get the attributes of a variable, MMS defines the GetVariableAccessAttributes service. A Read request specifies the name of one variable. This service does not allow to read the attributes of multiple independent variables.

The response sent by the server indicates whether the service was correctly executed or not. It returns the MMS Deletable, Address and Type Specification attributes of the variable.

This service is essentially used to obtain the type description of a variable. It is useful to check the actual definition of a variable in a manufacturing device.

4.4.5 Reading variables

To get the value of one or multiple variables, MMS defines the Read service. A Read request specifies the names of one or more variables and, for complex variables, an alternate access. Note that this last parameter is optional.

The response sent by the server indicates whether the service was correctly executed or not for each of the variables specified in the request. A Read request for a list of variables is equivalent to a series of Read requests for each of the variables in the list, in the sense that failure to read one of the variables does not affect the reading of other variables in this list.

For arrays and structures however, partial reading is not allowed: either all elements mentioned are read or the operation fails.

4.4.6 Writing variables

To modify the value of one or multiple variables, MMS defines the Write service. A Write request specifies the names of one or more variables and, for complex variables, an alternate access. Note that this last parameter is optional.

The response sent by the server indicates whether the service was correctly executed or not for each of the variables specified in the request. A Write request for a list of variables is equivalent to a series of Write requests for each of the variables in the list, in the sense that failure to write one of the variables does not affect the writing of other variables in this list.

For arrays and structures however, partial writing is not allowed: either all elements mentioned are written or the operation fails.

4.5 Summary

This chapter introduced the most important classes of MMS objects, used in the majority of applications: Domains, Program Invocations and Named Variables. These objects represent physical and logical resources of real manufacturing devices. For each class, we described the main attributes and services used in manufacturing applications. The notions of MMS name and association were defined in this chapter. Chapter 5 presents the objects and services available in three industrial devices while Chapter 6 discusses the use of these objects in a real manufacturing application.

5. MMS in Real Manufacturing Devices

5.1 Introduction

Chapter 4 introduced the Manufacturing Message Specification (MMS) as an application layer protocol for communication with manufacturing devices. MMS is defined independently of the device on which it is used. Indeed, MMS was designed to be applicable to a wide spectrum of industrial devices, from numerical controllers (NCs) of machines to Programmable Logic Controllers (PLCs). This approach is characterised by the introduction of the VMD (Virtual Manufacturing Device) abstraction, which is central to MMS.

In this chapter, we examine three examples of real manufacturing devices which are able to communicate with MMS. These are the Siemens SIMATIC S5 family of PLCs, the Siemens SINUMERIK line of numerical controllers and the Télémécanique TSX 47 line of PLCs. We shall see in particular in the case of the NC, how application-specific objects can be modelled using MMS objects and with which MMS services they can be accessed.

The principle when implementing MMS in a real manufacturing device is to off load the device from communications tasks such as encoding or decoding messages at the MMS level. For this reason, the MMS provider is practically always running on a separate board that is responsible for handling communications tasks of layers 1 to 7. In the Siemens families of manufacturing devices, these communication boards are called CP (Communications Processor). The rule in current implementations is that one such board implements one and only one Virtual Manufacturing Device (VMD), i.e. one MMS abstraction of a real manufacturing device. In the following sections, we will often mention the term "server" which means in this context the piece of software that implements the VMD abstraction and that is able to handle MMS requests. This server is located on the CP board in Siemens devices.

5.2 The Siemens MMS implementation for SIMATIC PLCs

The Siemens SIMATIC S5 family of Programmable Logic Controllers (PLCs) is used in a wide spectrum of applications such as process control, manufacturing automation, etc. These PLCs can be equipped with communication boards that implement the MMS protocol. These communication boards can give access to a maximum of four CPUs in the PLC and support up to 32 concurrent associations. Multiple communication boards can be plugged in the rack of a PLC.

5.2.1 VMD related services

With the Status service, a client can request the status of the VMD associated to a PLC. This status is composed of two pieces of information, one giving the physical status of the PLC, the other indicating its logical status. The PLC can also spontaneously send status notifications when significant state changes occur. This is done with the UnsolicitedStatus service.

The GetNameList service is handled by the server without intervention of the PLC user program. It may provide a client with names of the existing variables, domains and program invocations that can be accessed through MMS. In the same way, the server automatically responds to requests of a client concerning the identity of the PLC (Identify service).

5.2.2 Accessing variables in the PLC

An S5 PLC can act both as a client and as a server to read and write variables. These variables are accessed by name and can have one of the types listed in Table 5.1. When acting as a server, the PLC can provide a client with information on variables, for example the type of the variable.

Type	Meaning
BO	Boolean
IN	integer 8, 16 or 32
UN	unsigned 8, 16, 32
FP	floating point
TI	time of day, 4 octets
TD	time of day, 8 bytes
BS	bit string
OS	octet string
VS	visible string

Table 5.1: Data types supported by the CP 1473

Variables can be simple (booleans, integers, etc.) but also complex with arrays and structures. In the case of complex variables, a limit is placed on the nesting level which describes the complexity of the variable in terms of levels. This limit is set to two, which means that the following types are supported:

- arrays and structures of simple variables
- bidimensional arrays of simple variables
- arrays of structures containing simple variables
- structures containing arrays or structures of simple variables

Figure 5.2 gives an example of complex variables that can be stored on a S5 PLC and accessed by MMS through the Read and Write services. Note that the PLC user program can spontaneously send Information Reports containing values of simple or complex variables. This report of variable values can be done periodically or at particular conditions that are specific to the application.

```
int numbers[10];

struct {char name[8]; int temperature; int maxlimit};
```

Figure 5.2: Examples of data structures accessible by MMS in the Siemens Simatic S5 PLC

5.2.3 Managing domains in the PLC

In the MMS model, a domain may contain a program, a part of a program or data. Domains can be predefined or dynamically downloaded. In a SIMATIC S5 PLC, both kinds of domains can exist. These domains represent blocks used by the PLC for normal operation. Examples are:

- organisation blocks
- program blocks
- sequential blocks
- function blocks
- data blocks

A domain is composed of one or more blocks and the composition of a domain is decided by the user. In practice, very often domains contain one or two types of blocks so that modification of a production activity does not require the download of a whole set of blocks but only a transfer of those blocks that need to be modified.

Dynamic domains can be used to download on the PLC a STEP 5 program that has been produced on a programming station. Up to eight such domains can be managed by the server on the PLC. Selection of the CPU inside the PLC on which a domain will be transferred is done with the ListOfCapabilities parameter of the InitiateDownloadSequence service. Domains can be downloaded or uploaded directly from a cell controller, or the latter can instruct the PLC to load or store a domain on/from a file server. The PLC can provide information on one of its domains using the MMS GetDomainAttributes service.

5.2.4 Managing programs on the PLC

In a Siemens S5 PLC, two MMS Program Invocations (PI) can be supported. The first PI is static and used to start or stop the PLC. The second PI is used to control execution of a user application program running on one of the PLC CPUs. For this second type of PI, all MMS PI services can be used, namely Start, Stop, Resume, Reset, Kill. This PI can be compiled with the CreateProgramInvocation service from one or more of the 8 dynamic domains available on the PLC but this service is responded to automatically by the CP without intervention of the user program. The same is true for the DeleteProgramInvocation. The characteristics of the PI can be returned to a client invoking the GetProgramInvocationAttributes service. The state of the PI is stored in the server so that it can filter incoming requests and automatically respond to incorrect ones without intervention of the user program. For this scheme to work, it is necessary for the user program to inform the server of all state changes occurring so that the state variable in the server always reflect (with some short delay) the real state of the PI.

5.3 The Siemens MMS implementation for SINUMERIK NCs

SINUMERIK is the name of a family of numerical controllers developed by Siemens for manufacturing applications. Two models of the SINUMERIK family may be connected to a MAP or CNMA network using a specialised communications processor. The description of the NC given in the following sections is largely based on the NC Companion Standard to MMS that is implemented on the Siemens MAP interface. In particular, the names of variables used to store machine data has been standardised in the companion standard.

5.3.1 Structure of the NC

A SINUMERIK NC is composed of multiple processors, one of them being a PLC. The CP 1476 is a communications processor that implements the MAP and CNMA profiles, in particular the MMS. It is responsible for all communication functions, establishing associations, performing MMS services, releasing associations. Both the DIS and IS versions of MMS are supported by the CP.

NC programs, tool and machine data are stored on the NC CPU while all other data are stored on the PLC.

5.3.2 Accessing variables

In the Siemens NC-MMS, the NC can act both as client and as server for variable access. In both cases, variables are read or written by specifying their name and not their address. As a client, the NC can read and write variables in a remote device but one variable at a time. Variables accessed in the remote device can have any of the three defined scopes: vmd-specific, domain-specific, association-specific.

As a server, the NC can answer to Read and Write requests and can in some well defined conditions send Information Reports. The scope of variables managed as a server includes vmd-specific and association-specific. The types supported are the same as those of a SIMATIC PLC (see Table 5.1). The variables that are accessed by MMS are stored in the PLC part of the NC. To be accessible, these variables must be declared to a tool that will generate the appropriate management information for the server.

Variable services are used to read and write tool data or zero offsets from/to the tool management system of the NC. Additional user defined variables may be used to solve application specific communication problems, for example to do the synchronisation with a transport system.

5.3.3 Managing domains

The server allows data to be stored in a number of domains that can be classified in the following types:

- *program domains* contain NC programs and are downloaded using MMS domain services.
- *data domains* are predefined and contain information on the machine and tools used.
- *device domains* group together information on NC subsystems such as tool changer or rotary table. Up to eight device domains can be supported by the CP.
- a *trace domain* records actions performed by the CP 1476 to diagnose potential errors.

Data domains are predefined and contain information on the machine and tools used, for example configuration data, active zero offsets, NC and PLC data.

Dynamic domains of the SINUMERIK NC may be downloaded or uploaded and the NC can be asked by a cell controller to load or store the content of domains in or out of the NC.

Whatever the type of domain, a client may receive information on existing domains using the GetDomainAttributes service while the GetNameList service indicates which domains are present in the NC.

5.3.4 Controlling programs

The Siemens NC VMD supports the notion of Program Invocations but these are not directly mapped onto NC programs. For reasons of safety, PI service requests sent to the NC are forwarded to the PLC and not to the CPU executing the program. The user program running on the PLC then passes the request to the appropriate CPU. To specify the subsystem on which a PI will run, a client gives the device name in the CreateProgramInvocation request together with the name of the PI. All subsequent requests can then be forwarded to the appropriate device on the basis of the PI name.

Rather than the client periodically polling the state of an executing program to determine when it has completed, the NC is able to send an EventNotification to the client provided that the client has requested this mechanism when the Program Invocation was created .

5.3.5 Alarms

Significant events detected by the PLC or the NC may be reported to the cell controller using the MMS EventNotification service. In the case that this service is not supported, the events may be reported using MMS InformationReports. Alarms coming from the NC almost always stop the machine while alarms generated by the PLC do not cause the machine to stop. Event notifications may also contain supplementary information, provided by Siemens, for example user interrupts, operation messages and resume messages.

5.4 The Télémécanique MMS implementation for TSX 47 PLCs

The Télémécanique TSX 7 family of Programmable Logic Controllers (PLCs) can be equipped with communication boards that implement the MMS protocol. As for the Siemens PLCs, the TSX communication interfaces support the whole MAP protocol stack, including the MMS provider. A server is implemented on the board to deal with all services that do not require processing by the PLC itself. The MMS functionality supported by Télémécanique is essentially equivalent to that provided by Siemens: VMD, variable, domain and program invocation management. A comparison at the end of the chapter indicates the similarities and differences between the PLCs of the two manufacturers.

5.4.1 VMD related services

The PLC can respond to Status requests, providing the physical and logical status of the PLC or can spontaneously send an UnsolicitedStatus notification to another device. As a client, the PLC can request the status of a remote device and receive from the remote device an UnsolicitedStatus.

The GetNameList service is handled by the server without intervention of the PLC user program. It may provide a client with information on the existing variables, domains and program invocations that can be accessed through MMS. In the same way, the server automatically responds to requests of a client concerning the identity of the PLC (Identify service).

5.4.2 Accessing variables in the PLC

A TSX PLC can act both as a client and as a server to read and write variables. These variables are accessed by name or by address and can have one of the types listed in Table 5.2. As a server, the PLC can respond to Read and Write requests concerning up to 16 different variables while as a client, the PLC is allowed to access a maximum of 4 variables in a remote device. Both named and unnamed variables are supported by the server. Only two VMD-specific variables are defined, namely Date_Time and System_Bit. All other variables are either domain-specific or application association-specific. Unnamed variables are accessed by symbolic addresses well known to PLC programmers, for example W10 for the 10th word or B7 for the 7th bit of the application domain.

Type	**Meaning**
System_Bit	Boolean
Internal_Bit	Boolean
Internal_Bit_Word	Boolean
Internal_Word	integer16, unsigned16, bitstring16, octetstring2
Internal_Double_Word	integer32, unsigned32, floating point
Constant_Word	as Internal_Word
Constant_Double_Word	integer32, unsigned32, floating point
System Word	as Internal_Word
Date_Time	Generalized_Time
String of Internal Words	array, structure, octetstring2N, visible string2N
String of Internal_Bits	bit string N, array of boolean

Table 5.2: Types supported by a Télémécanique PLC

Variables can be simple (booleans, integers, etc.) but also complex with arrays and structures. In the case of complex variables, a limit of two is placed on the nesting level, which means that the following types are supported:

- arrays and structures of simple variables
- bidimensional arrays of simple variables
- arrays of structures containing simple variables
- structures containing arrays or structures of simple variables

For complex variables, it is possible to use the alternate access mechanism that allows the selection of specific elements of an array or particular components of a structure.

5.4.3 Managing domains in the PLC

In the MMS model, a domain may contain a program, a part of a program or data. In a Télémécanique TSX 47 PLC, a single domain can exist. This domain can be downloaded, uploaded and deleted via MMS services. It contains both the user application program and the data.

5.4.4 Managing programs on the PLC

In a Télémécanique TSX 47 PLC, a single Program Invocation (PI) can be supported. This PI is used to control execution of a user application program running on the PLC. All MMS PI services - except Kill - can be used, namely Start, Stop, Resume, Reset. No argument can be passed to the PI at start time. The PI can be compiled with the CreateProgramInvocation service and removed from the PLC by a DeleteProgramInvocation. The characteristics of the PI can be returned to a client invoking the GetProgramInvocationAttributes service.

5.5 Comparison of the two PLC MMS implementations

Table 5.4 summarises the similarities and differences between the Siemens and the Télémécanique PLC. As can be seen from this table, the PLCs offer approximately the same functionality as servers. As clients, the Siemens PLC is able to remotely control the execution of programs while the Télémécanique does not have this capability.

Feature	Siemens S5 PLC	Télémécanique TSX 47 PLC
Max. number of associations	32	32
MMS versions supported	DIS, IS	DIS, IS
Domains	8	1
Program invocations	2	1
PI services as server	all	all except Kill
PI services as client	all	none
Variable support	VNAM, VADR	VNAM, VADR
Read, Write (Client)	1 variable	4 variables
Read, Write (Server)	limit is maximum PDU size	16 variables
Arrays and structures	yes	yes
Max. nesting level	2	2
Alternate access	yes	yes
Event Notifications	no	no

Table 5.4: Comparison of the Siemens and Télémécanique MMS implementations for PLC

5.6 Summary

In this chapter, we examined three examples of real manufacturing devices that are able to communicate with MMS. These are the Siemens SIMATIC S5 family of PLCs, the Siemens SINUMERIK line of numerical controllers and the Télémécanique TSX 47 line of PLCs. We have seen in particular in the case of the NC, how application-specific objects can be modelled using MMS objects and with which MMS services they can be accessed. A comparison was made between the Siemens and Télémécanique MMS implementations for PLCs which roughly provide the same functionality. Chapter 6 describes the use of MMS in a real manufacturing application and illustrates how MMS objects available in PLCs and NCs can be used for the modelling of this application.

6. Use of MMS in a Real Manufacturing Application

This chapter illustrates the use of MMS in a real manufacturing application. It considers a modern production facility at Mercedes-Benz, used for the manufacturing of car engines. The example scenarios described are considered helpful to the general understanding of MMS in a wide range of typical manufacturing applications.

In this chapter, the reader will find an overview of the manufacturing facility and a detailed description of which MMS objects are used to control and monitor the operation of the Flexible Manufacturing System (FMS).

We start with a short overview of the manufacturing facility and then describe the information model and typical scenarios for a rail guided vehicle and the machining centres. The information model describes the variables that are used to store information concerning the parts, tools and orders handled by the FMS. The scenarios illustrate the MMS services that are used by the FMS controller to access these pieces of information and to request execution of particular operations by the manufacturing devices.

6.1. Overview

The manufacturing facility considered in this chapter consists of three machining centres, one automated and two manual clamping workstations for the provision of parts, a storage and a rail guided vehicle (RGV) to supply the machining centres and the workstations with pallets, as shown in Figure 6.1. Parts are fixed on pallets that are used to transport them between the different nodes of the transport system.

An FMS controller (referred to as the Controller in the following) supervises the operation of this FMS. It provides machining centres with the data necessary for their operation, requests missions to the RGV controller and monitors the status of all manufacturing devices.

The following sections describe how MMS can be used to control and monitor a Rail Guided Vehicle and a Machining Centre in the context of the manufacturing operation outlined above.

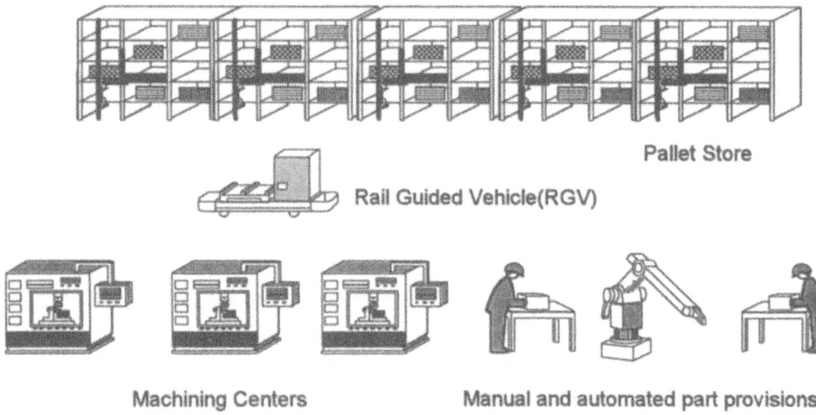

Figure 6.1: The Mercedes-Benz manufacturing cell

6.2. Rail guided vehicle (RGV)

The rail guided vehicle is controlled by a PLC that acts as an MMS server. The PLC provides status information about all available material flow nodes of the transport system and gets transport orders from the FMS controller. Material flow nodes are places where the vehicle loads or unloads pallets, which contain the parts or tools. There are three different types of transport orders. The vehicle may "Fetch pallet <palletId> from node <nodeId>", "Put pallet <palletId> to node <nodeId>" and "Move to node <nodeId>". The PLC is required to check the feasibility of the received transport order. This includes checking the current state of the destination and/or source node. A material flow node of the storage may be either empty or full. Material flow nodes of the other stations may be empty, reserved, occupied or unknown.

The externally accessible information necessary for the control of the RGV is mapped to MMS Named Variables which reside and are predefined in the PLC.

6.2.1. RGV state diagram

A state diagram describes the behaviour of the RGV as visible from the FMS Controller. Figure 6.2 shows the possible state transitions whilst Table 6.1 indicates the states in which the RGV may be.

State	Description
Idle without pallet	There is no pallet on the RGV. The RGV is waiting for a transport order
Driving without pallet	The RGV has started the execution of a transport order and is on the way to the node.
Pick up a pallet	The RGV has arrived at the node and has started to pick up the pallet.
Idle with pallet	The pick up procedure was finished successfully and the RGV is waiting for a transport order.
Driving with pallet	The RGV has received a transport order and is driving to the destination node.
Give up a pallet	The RGV has arrived at the destination node and has started the give up procedure.

Table 6.1: States in which the RGV may exist.

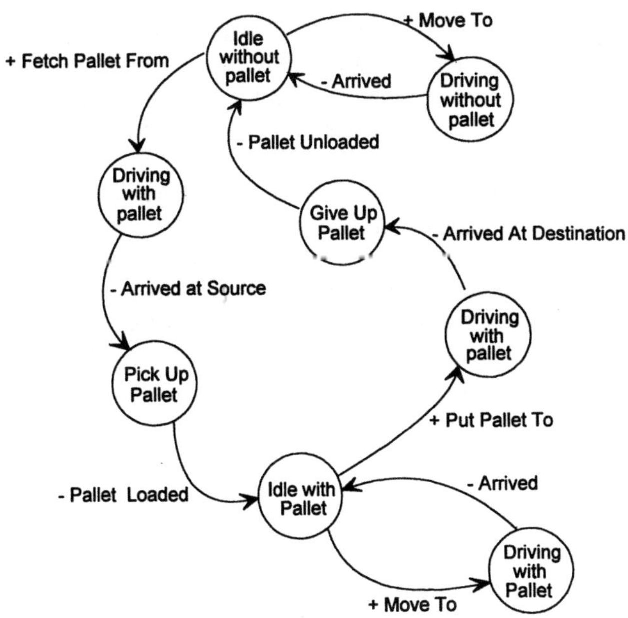

Figure 6.2: RGV State Diagram

6.2.2 Transportation of a pallet from a clamping station to the pallet store

The purpose of this operation is to move a pallet from a clamping node to the pallet store. The scenario is illustrated in Figure 6.3.

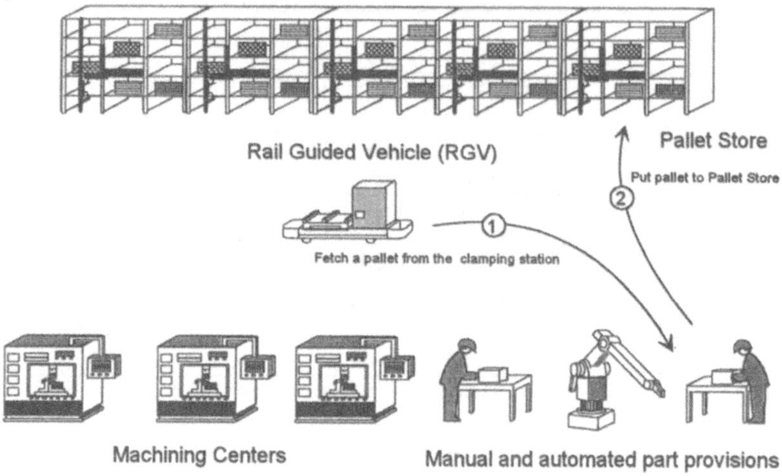

Figure 6.3: Transportation of a pallet from a clamping node to the pallet store.

The operation is supervised by the FMS controller, which gets the necessary information from and sends the transport orders to the PLC. The sequence of exchanged MMS messages is shown in Figure 6.4. The Controller initiates the scenario after the PLC has notified the state change of the clamping node to the Controller .

The Controller sends a transport order of type "Fetch a pallet from a node" to the RGV. If the RGV accepts the transport order, the Controller requests the starting of the transportation. The RGV notifies the Controller, when it has arrived at the source node and after it has picked up the pallet. After that, the RGV waits for the next transport order. The Controller then sends a transport order of type "Put pallet <palletId> to node <nodeId>". After execution of this command, the pallet is stored in the pallet store.

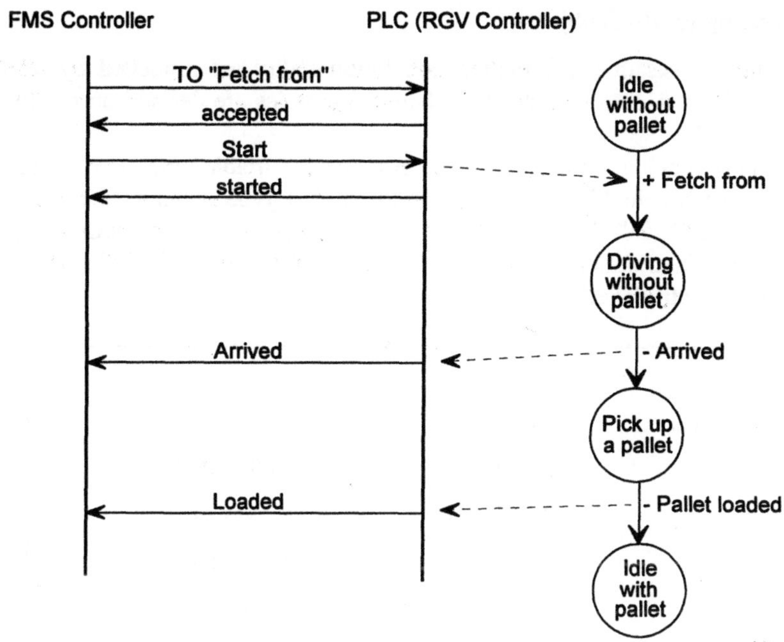

TO Transport order
FMS Flexible Manufacturing System
RGV Rail Guided vehicle
PLC Programmable Logic Control

Figure 6.4: Sequence of messages for the scenario of Figure 6.3

6.2.3 Transportation of a pallet from the pallet store to a machining centre

The RGV supplies the machining centres with parts fixed on pallets. Pallets may come either from the pallet store or from one of the three clamping stations. Before the FMS controller requests the transportation of a pallet to the machining centre, it has to check whether the node associated to the machining centre is used or not. If the node is not used, the Controller reserves this node for the transportation of the pallet. Then, it sends a transport order of type "Fetch pallet <PalletId> From <pallet store node>" to the RGV and instructs the RGV to put the pallet on the node corresponding to the machining centre.

6.2.4 Mapping to MMS objects

The externally accessible information and functionality are modelled by MMS Named Variables. A set of predefined Named Variables are defined on the PLC driving the RGV.

A transport order is represented by a structured variable composed of three elements, as shown in Figure 6.5. The element TO_type contains the type of transport order ("Fetch", "Put" or "Move"). The element NodeName indicates the name of the destination node. The identity of the pallet to move is defined by the value of the element PalletId.

Name	**TO**	
Type Description	struct {	
	TO_type	Octet String[6]
	NodeName	Octet String[3]
	PalletId	Octet String[22]
	};	
MMS Deletable	FALSE	

Figure 6.5: Named variable representing a transport order

The TO_start Named Variable is used to initiate the processing of a transport order specified by the variable TO, as shown in Figure 6.6. It can be regarded as an On/Off switch. To start the execution of the transport order the Controller changes the value from FALSE to TRUE. After the completion of the transport, the value is changed from TRUE to FALSE by the PLC by local action.

Name	**TO_start**
Type Description	Boolean (TRUE,FALSE)
MMS Deletable	FALSE

Figure 6.6: Named variable used to start a transport order

The ShuttleReport (as shown in Figure 6.7) is a structured variable whose value is systematically sent by the PLC after each state transition defined by the RGV state diagram, shown in Figure 6.2. It allows the FMS controller to check the operation of the RGV and identify the state in which the device is currently.

Name	**ShuttleReport**	
Type Description	struct {	
	TO_type	Octet String[6]
	TO_PalletId	Octet String[22]
	State	Integer16
	ErrorCode	Integer16
	Store_Node	Octet String[3]
	Store_PalletId	Octet String[22]
	};	
MMS Deletable	FALSE	

Figure 6.7: Named variable for ShuttleReport

The nodes of the material flow system can be in one of the following states: empty, reserved, occupied or unknown. There is one MMS Named Variable of the type represented in Figure 6.8 for each transport node in the material flow system. This variable indicates the state of the node as well as the PalletId if a pallet is stored on the node.

Name	**<nodeId>**	
Type Description	struct {	
	state	Octet String[6]
	PalletId	Octet String[22]
	};	
MMS Deletable	FALSE	

Figure 6.8: Named variable describing a node of the transport system

6.2.5 MMS services offered by the RGV

Given the fact that the RGV is modelled by MMS variables only, MMS services Read, Write and InformationReport are sufficient for communication between the FMS controller and the PLC controlling the RGV. Additional services, such as Initiate and Conclude are necessary for the management of MMS associations and are not considered in this chapter.

Figure 6.9 illustrates the mapping onto MMS services of the scenario described in Figure 6.4. A transport order is specified to the PLC using the Write service. The move is started by writing the TO_start variable. The PLC notifies state changes of the RGV and involved nodes by using the MMS InformationReport service.

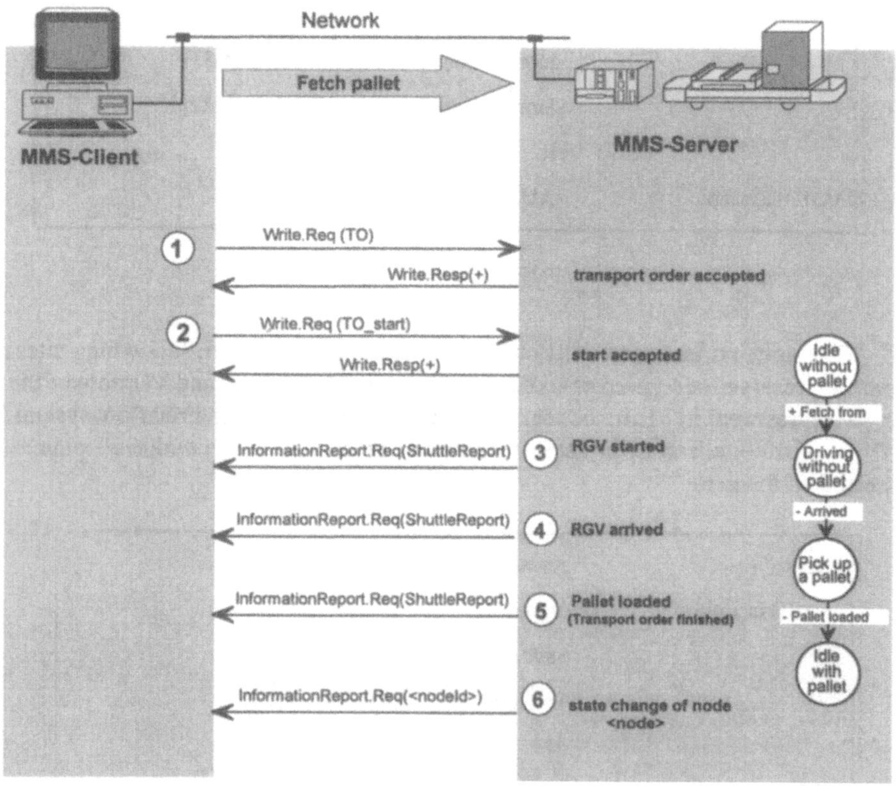

Figure 6.9: Sequence of MMS services used for fetching a pallet

First of all, the FMS controller has to check the feasibility of the transport order. The transport order may be completed successfully if and only if the destination node <nodeId> is empty. Therefore the Controller reads the variable <nodeId> in order to get the actual state. If the node is empty, the Controller reserves the node. This is done by setting the value of the element State of the variable <nodeId> to "reserved". After these preparation activities the rest of the scenario of Figure 6.10 is identical to the scenario illustrated in Figure 6.9.

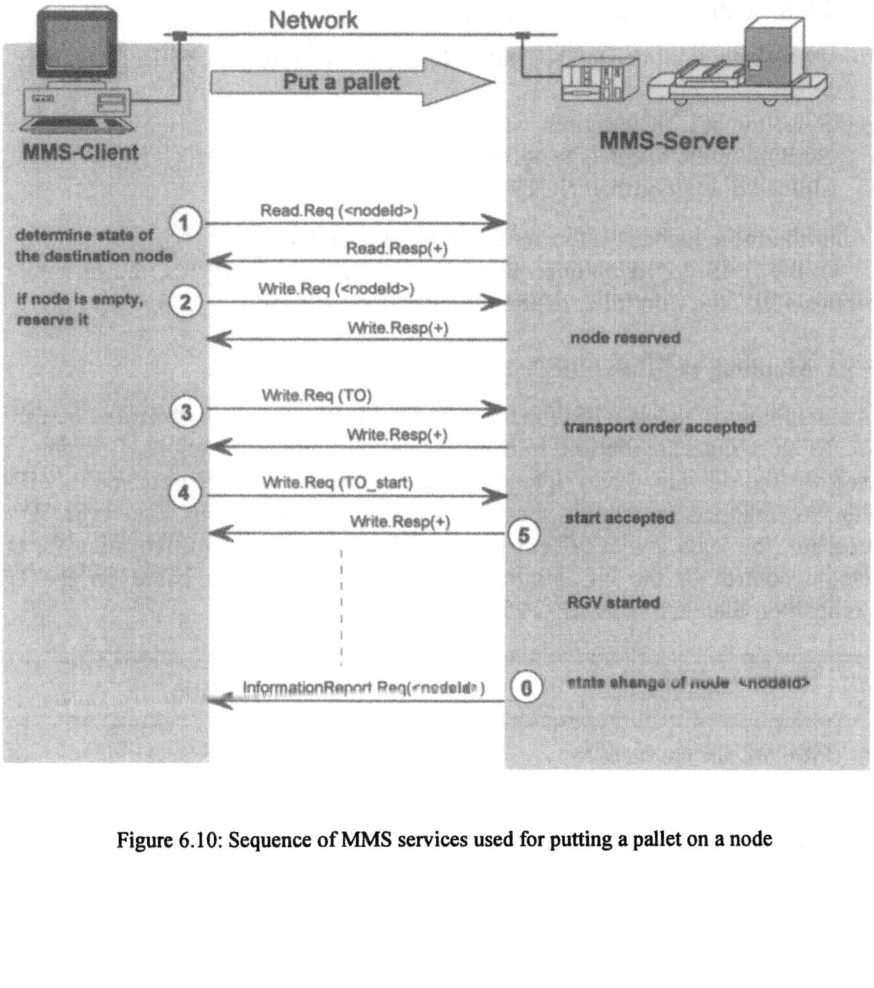

Figure 6.10: Sequence of MMS services used for putting a pallet on a node

6.3. Machining centre

The main manufacturing devices of the FMS are the machining centres. The FMS controller supplies the machining centres with all manufacturing resources as well as the programs and data necessary for the automated manufacturing of parts. In the following sections a description is given of the main communications aspects.

6.3.1. Overview of the information flow

The FMS Controller is mainly responsible for the:

- Download of all needed NC data (NC programs, tool data, pallet offset data, etc.)
- Selection of a NC program
- Starting of the selected NC program
- Upload of all modified NC data.

Furthermore it controls the exchange of parts and tools.

As the FMS is programmed to react to asynchronous events, the machining centres notify the Controller of state changes and failures.

6.3.2. Mapping to MMS objects and services

The mapping of NC-specific functionality to MMS services is indicated in Table 6.2. NC programs are mapped to Domains and Program Invocations. Domains are used for the download and upload of NC programs. Tool data and pallet offset data are mapped to domain-specific variables. All these NC data types (NC program, tool data, pallet offset data) are linked to a Program Invocation for the remote control of the NC program. The mapping used is based on the NC Companion Standard [ISO/IEC 9506-4].

Functionality	MMS services
Download and upload of NC programs	Domain download and upload services
Transfer of tool data and pallet offset data	Variable access services
Execution control of NC programs	Program Invocation services
Notification of the FMS Controller	Unsolicited Status and Information Report

Table 6.2: Mapping of functionality onto MMS services

The FMS controller manages the download and upload of all NC programs, which are necessary for the manufacturing of parts and for the exchange of tools. NC programs which are not needed anymore are deleted by the Controller. Program Invocations are used for the remote control of NC programs. An NC program needs a set of tools, which must be available on the machine tool. As the tool data and fixture offset data are not stored in the NC program, they must be linked to the appropriate NC program. The Controller provides all needed tool data and information necessary for this binding procedure.

6.3.3 NC-CS models used in the example manufacturing facility

The NC Companion Standard (NC-CS) introduces six models to describe the capabilities and resources of a NC system. These can be seen in Table 6.3. Most of them are used in this example manufacturing facility. The following sections give a short description of the most important models.

Model	Note
• NC **System State Model**	Represents the overall state and the operation mode of a NC
• NC **Device Control Model**	Used for the control of NC programs
• NC **Information Model**	Provides basic information on the NC
• NC **Offset Model**	No relevance to the description of the externally visible behaviour.
• NC **Alarm Model**	Modified for this example manufacturing facility.
• NC **Data Store Model**	Describes the current state of the NC data store

Table 6.3: NC models used in the example manufacturing facility

6.3.4 NC system state model

The NC System State Model describes the overall state and operation mode of an NC system.

The overall state may take on the values NORMAL PRODUCTION or LOCAL ACTION REQUIRED and is mapped to the attributes LogicalStatus and PhysicalStatus of the VMD object.
The operation mode of the NC is mapped to the MMS Named Variable N_Remote of type Boolean. To switch from local to remote control, local action by the operator is required. This transition causes the value of the Named Variable N_Remote to change from FALSE to TRUE. The Controller is not allowed to trigger this state transition.

The transition from remote to local control may be initiated by local means (operator intervention or safety violation) or by the FMS controller sending a MMS Write Request to set the value of N_Remote from TRUE to FALSE.

The NC may notify the FMS Controller of changes of the operation mode or overall state by sending an InformationReport or an UnsolicitedStatus. The use of these MMS services is described and illustrated in chapter 4.

The relationship between the local/remote state and the VMD Logical and Physical status is shown in Table 6.4. It should be noted that this relationship is used in the example manufacturing facility and not defined by the NC-CS.

Mode Of Operation	N_Remote	VMD physical status	VMD logical status
Remote	TRUE	Operational	State Changes Allowed
Local	FALSE	Operational	Any
Needs Commissioning	FALSE	Inoperable	Limited Services Permitted

Table 6.4: Relationship between operation mode, physical and logical status of the VMD.

6.3.5 NC device control model

The NC-CS uses an NC device control model permitting the control of one or more independent devices in a single NC system. The model is defined independently of the type of device. Basically the device control model incorporates an image of each device, including information about the physical resources, the current state and the operating parameters of the device. Additionally there are part programs, data etc. Execution of the part program is represented by the controlling process linked to the image of the device, the part program, data etc. and described by a state model. The way in which a system is delineated in devices is arbitrary and implementation-specific. It is possible to link a single device to several controlling processes or several devices to a single controlling process.

In the NC-CS, the image of every device is mapped onto a device domain. The device domain contains Named Variables describing the state and parameters of the device. The NC-CS defines standardised names for Named Variables commonly used with devices. The standardised prefix for device domains is N_DEV_.

The part program is stored in program domains whose names begin with the standardised prefix N_PRG_.

A process information domain holds information about the state of a controlling process, e.g. the program line that is currently being executed. Its name begins with the standardised prefix N_PID_.

A controlling process controls the execution of the user program. Linked to the Controlling Process are at least a device domain, a program domain and a process information domain. It is possible to link several devices acting in a co-ordinated fashion to a single Controlling Process. It is also possible to define several independent Controlling Processes in a single system and to link a single device to several controlling processes at the same time. The standard does not make any restriction on the usage of devices and controlling processes. The actual structure is defined by the device manufacturers.

The controlling process is mapped on the Program Invocation object as defined in the MMS-Core [ISO 9506-1] without adding any additional attributes or procedures. The Controlling Process state diagram is directly mapped to the Program Invocation state diagram. There is no standardised name for the Controlling Process.

A simplified structure of the VMD of an NC-system with two devices acting in a co-ordinated way is shown in Figure 6.11.

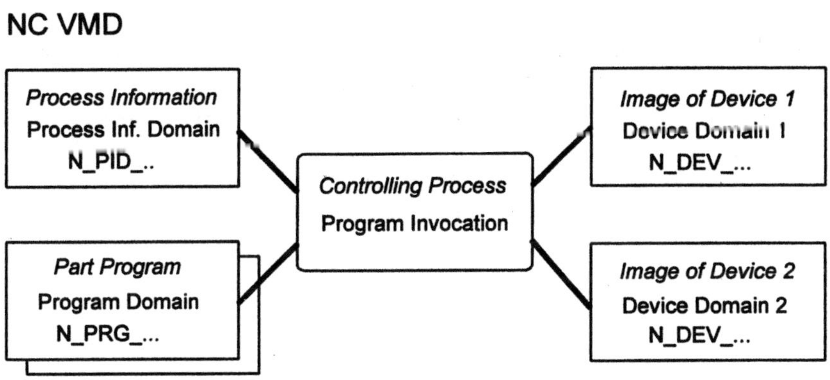

Figure 6.11: VMD of an NC composed of two devices

The NC-CS defines standardised names for Variables that are specific to a device domain. Some of these variables are listed in Table 6.5.

Standardised Named Variables	Definition
N_NumAxes	Number of Machine Axes
N_NumSpindles	Number of Spindles
N_FRL	Array of floating point values for Feed Rate Limits for each axis
N_FRO_ON	Boolean variable to switch remote Feed Rate Override on and off
N_FRO	Feed Rate Override for the whole device
N_RPO_ON	Boolean variable to switch Rapid Override on
N_RPO	Rapid Override for the whole device
N_SSL	Array of floating point values for Spindle Speed Limits for each spindle
N_SSO_ON	Boolean variable to switch Spindle Speed Override on and off
N_SSO	Spindle Speed override for the whole device

Table 6.5: Named Variables describing a device

Note that these Named Variables are optional and therefore may or may not be implemented by the vendor of the device controller.

The example manufacturing facility makes use of the defined prefix N_PRG_ for domains which contain the NC part programs and for the Controlling PIs. Thus, the domain and the associated PI have the same name. For example, Domain N_PRG_12345678 is linked to the Controlling PI N_PRG_12345678. PIs are used for the manufacturing of parts, for the exchange of tools and for the exchange of workpieces.

6.3.6 Alarm model

Alarms represent events of major importance to the FMS controller in the sense that they indicate abnormal conditions. The example manufacturing facility does not make use of the NC-CS Alarm Model which is rather complex. Instead, it relies on a simple mechanism. In the example manufacturing facility, an NC contains the Named Variable N_Alarm and notifies the occurrence of an alarm to the FMS Controller by sending the value of this variable in an Information Report.

The N_Alarm Named Variable has the structure defined in Table 6.6. The variable indicates the class and group to which this alarm belongs, its name and where the alarm occurred.

Name	**N_Alarm**	
Type Description	Class	VisibleString[80]
	Group	VisibleString[80]
	Name	Integer16
	Text	VisibleString[80]
	Location	VisibleString[80]
MMS Deletable	FALSE	

Table 6.6: Definition of the NC alarm variable

6.3.7 NC data store model

The FMS controller is responsible for the handling of the NC data store. One of the most important tasks of the Controller is to delete NC data that are no longer needed. To perform this, the Controller needs information about the actual state of the NC data store. The NC provides a set of domain-specific Named Variables, which reflect the capacity and current state of the data store. The name of the Domain is N_STORE_0. All objects are created and modified by the NC. The FMS Controller can only read the domain-specific Named Variables. These variables are listed in Table 6.7.

Name	Type
N_StoreType	VisibleString[32]
N_MaxStoreEntries	Unsigned32
N_MaxStoreCapacity	Unsigned32
N_CurrentEntries	Unsigned32
N_RemainingEntries	Unsigned32
N_CurrentCapacityUsed	Unsigned32
N_RemainingCapacity	Unsigned32

Table 6.7: Domain-specific variables describing the tool magazine

6.3.8 Tool management

Tool management is concerned with the management of all data describing tools used by machine tools. These data are stored in databases and transferred to the NCs by the FMS Controller. The NC Companion Standard defines a number of variables for the description of tool data. This Companion Standard is complex and covers more functionality than is needed in the pilot discussed here. The definitions given in this section correspond to the Heller flexible machining centres and are based on this standard. Machines supplied by other manufacturers may have a different representation of the same data, a situation which complicates the work of the FMS application developer.

All Named Variables which are defined for tool management are domain-specific. The name of the domain is N_TLD_0 as specified in the NC-CS. The domain is predefined and created by the set-up procedure of the NC system and contains the Named Variables N_TOOLMag_1, N_Tool_<ToolId>. All tool-specific Named Variables N_Tool_<ToolId> and N_Edge_<ToolId>_<EdgeId> are created by the NC-system by local means at the transfer of the tool data. This transfer is done before the tool exchange takes place.

Variable N_TOOLMag_1 describes the tool magazine in terms of the number of tools that are stored, the type of magazine and number of positions to hold the tools. This can be seen in Figure 6.12.

Name	**Domain: N_TLD_0;**	
	Variable: N_TOOLMag_1	
Type Description	struct {	
	StoredObjects	Unsigned8
	MagazineType	VisibleString[32]
	NumberOfPositions	VisibleString[32]
	};	
MMS Deletable	FALSE	

Figure 6.12: Named variable describing the tool magazine

Each tool is described by a domain-specific structured variable, the elements of which provide information on the tool state, location, edges, etc. as shown in Figure 6.13. These variables are stored in the same domain as the variable describing the tool magazine.

Name	**Domain: N_TLD_0;**	
	Variable: N_Tool_<ToolId>	
Type Description	struct {	
	ToolType	VisibleString[32]
	Location	VisibleString[32]
	Position	VisibleString[32]
	ToolEdges	Array Of VisibleString[32]
	ToolState	Unsigned8
	ToolDimensions	Array Of FloatingPoint
	....	
	};	
MMS Deletable	FALSE	

Figure 6.13: Named variable describing a tool

Name	**Domain: N_TLD_0;**	
	Variable: N_Edge_<ToolId>_<EdgeId>	
Type Description	struct {	
	ToolId	VisibleString[32]
	EdgeOffsets	Array Of FloatingPoint
	NominalEdgeLife	Unsigned32
	RemainingEdgeLife	Unsigned32
	....	
	};	
MMS Deletable	FALSE	

Figure 6.14: Named variable describing a tool edge

The tool edges are described in more detail by domain-specific structured variables of the type given in Figure 6.14. The ToolId references the tool this edge is part of. The edge offsets precisely indicate the relative position of the edge in the tool. The NominalEdgeLife and RemainingEdgeLife are used to assess how long the edge can be used. This data structure is the second example illustrating the use of floating points in MMS.

Finally, note that all these variables describing the magazine, tools and tool edges are stored in the same domain called N_TLD_0.

6.3.9 NC program management

The FMS controller must handle the download, the execution and upload of NC programs. Figure 6.15 illustrates the MMS services used for the NC program managament.

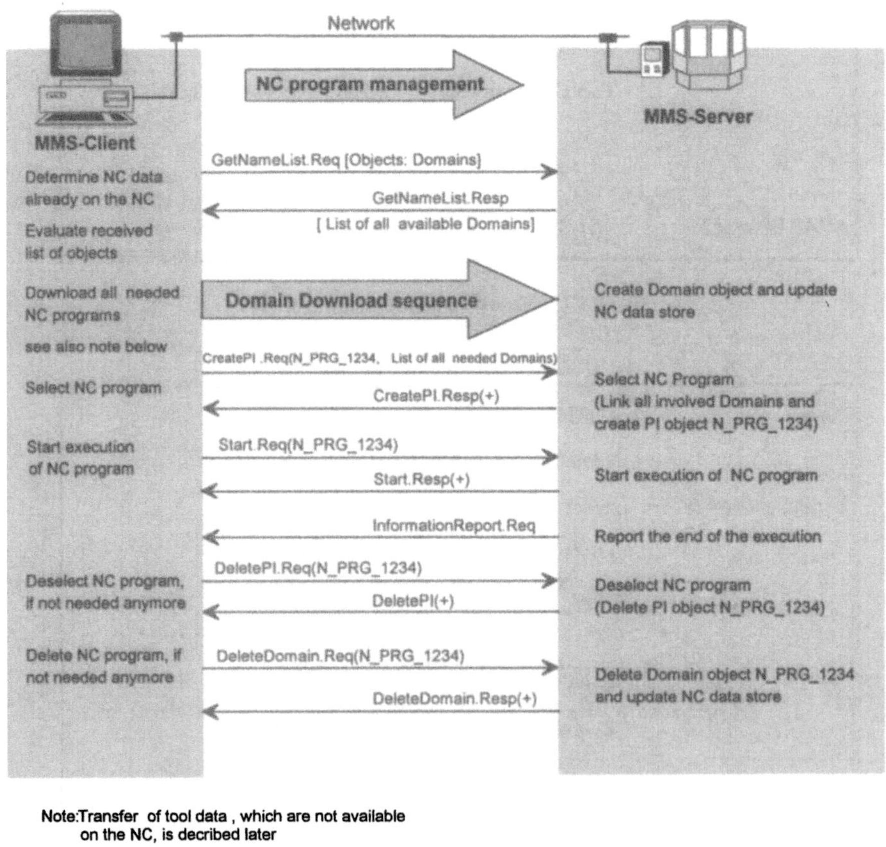

Figure 6.15: MMS services used for the NC program management

First, the Controller determines with the GetNameList service if the needed NC program is already available on the NC. If the NC program is not available on the NC, the Controller examines the state of the data store. If the result of the examination is positive, it initiates the download procedure. In addition, all needed NC data are transferred to the NC. Then, the FMS Controller selects the NC program and starts its execution. The NC notifies the Controller of the end of execution (InformationReport). The Controller deletes the NC program with the DeleteProgramInvocation service if it is not needed anymore.

6.3.10 Tool data transfer

Note that the NC data, which are needed for the current manufacturing order and which are not already available on the NC, are downloaded or transferred to the NC. This requires that the FMS Controller be able to determine which NC data are available on the NC. As most of the NC data are mapped onto Domains or domain-specific Named Variables, the FMS Controller may use the GetNameList service, which provides a list of MMS objects available on the NC. After evaluation of this list, the FMS Controller initiates the transfer of the required NC data as illustrated in Figure 6.16.

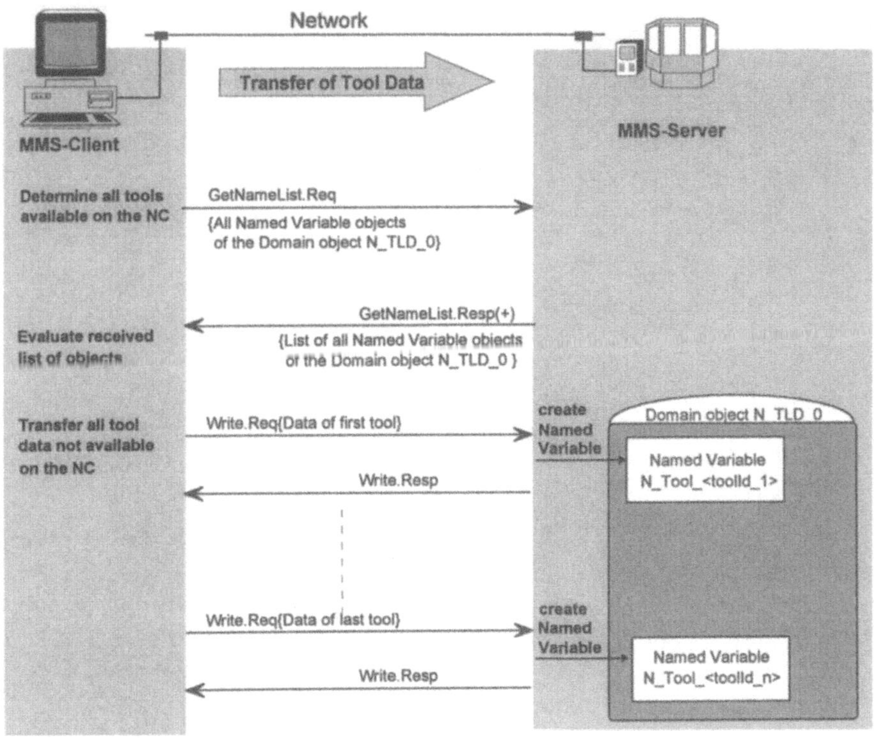

Figure 6.16: Transfer of tool data from the FMS Controller to the NC.

The NC notifies any change of the tool data to the FMS Controller by use of the InformationReport service. The FMS Controller must save all modified tool data.

6.4 Summary

This chapter illustrated the use of MMS in a real manufacturing application. It considered a modern production facility at Mercedes-Benz, used for the manufacturing of car engines. A short overview of the manufacturing facility was given and then the information model and typical scenarios for a rail guided vehicle and the machining centres were described. For the latter type of device, a simplified version of the MMS Companion Standard for NCs was used. Numerous figures illustrate the data structures describing physical resources of the manufacturing facility such as tool magazines and tools or describing messages such as transport orders. The reader should have a better understanding of the MMS modelling approach.

7. MMSI

This chapter presents the MMSI application programming interface. MMSI is the programming interface recommended in MAP to access the MMS protocol. It is assumed that the underlying operating system offers multitasking. The MAP specification defines an abstract interface that can then be particularised for specific languages. Currently, the only binding defined is for the C programming language. In practice, all existing MMSI implementations run on UNIX workstations. DOS and Windows applications use much simpler interfaces to MMS, an example of which is given in Chapter 8 with the MMS/DDE interface developed in ESPRIT project CCE-CNMA.

The main benefits of a standard Application Programming Interface (API) are discussed. The purpose, scope and properties of this interface are then presented. The chapter then introduces the three main classes of interface functions (sections 7.5 to 7.9). The more technical issues of function parameters, calling modes, Data Control Block management and object handles are presented in detail in sections 7.10 to 7.14.

The chapter includes numerous examples to give the reader a concrete idea of this interface. Readers interested in a general introduction can skip sections 7.10 to 7.14: these sections discuss in more detail different aspects of MMSI. They should however be of interest to readers involved in designing and programming MMS applications.

7.1 Introduction

The MMS protocol standardises the communication with manufacturing devices using an abstraction called Virtual Manufacturing Device (VMD). It specifies the message formats, the protocol rules to exchange messages and the behaviour expected from the VMD on receipt of a specific message. It guarantees that any device which conforms to the MMS standard can receive MMS messages from the network, decode them and perform the expected actions.

The MMS protocol does not specify how application programs interface with the MMS services: ISO standard protocols do not specify how each of the seven layers of the OSI model interfaces with its adjacent layer. MMS specifies the service that is to be provided at the application level but does not specify the user's view of the MMS service.

Vendors are free to define the interface between the user application program and the MMS service. The level of support offered by an **Application Programming Interface (API)** can vary widely between protocol implementations both in functionality and independence from the hardware platform and the operating system.

The main motivation for standardising an Application Programming Interface is to ensure the portability of application programs in an open environment with heterogeneous platforms and heterogeneous software and to minimise the training effort and development costs for new applications. The MAP 3.0 task force formed an API committee to develop a standard set of APIs for the different Application Service Elements (ASE). Three APIs have been defined: MMSI to access MMS, FTAMI to interface to the FTAM protocol and PCI, an interface to the ACSE and presentation protocols.

These APIs define a set of high-level language function calls and their arguments. They are designed according to the same model and share a set of common functions for event management, buffer management and connection management. They offer a unified and standardised interface to the different services of the application layer: MMS, FTAM and ACSE.

The purpose of this chapter is to present the MMSI interface, first in its principle, then in more details.

7.2 The main benefits of a standard API

There are numerous reasons to adopt a standard interface: reduction of costs and dependency on suppliers and protection of user investments.

One of the problems in a heterogeneous environment is the multiplicity of interfaces to use, for example to communicate with manufacturing devices. Even in the case of a standard protocol such as MMS, computers used for cell control often offer a proprietary interface. A program written for one machine of supplier 'A' will not run, without major porting effort on a machine of supplier 'B'. For a user, this situation is absurd, as the program is expected to do exactly the same things on the two computers. A standard interface allows a user to port a given program with a minimum of modifications from one machine to another. MMSI was designed for this purpose.

Training costs, which can be high if a large number of people are to be trained, are also reduced as programmers have to deal with one and only one interface, instead of multiple interfaces.

In some cases, when a new application has to be developed, it is possible to reuse a large part of an existing application. This results not only in a reduction of the programming effort but also in the design and testing of the software.

Another important advantage of standard interfaces is the independence gained with respect to suppliers. 'A' program developed initially on a machine from supplier 'A' can be ported on the machine of supplier 'B' if 'A' is no longer adopted as supplier by the user.

7.3 Purpose and scope of MMSI

The role of the MMS protocol is to build and unbuild the messages corresponding to the MMS service requests and responses and to do this according to the MMS protocol rules.

The purpose of MMSI is to present network data in a standard and useful format for the user program, see Figure 7.1. It defines a set of *function calls*, types and parameters for accessing MMS. These definitions are available in abstract form, independent of the programming language and in the C programming language. The existence of the abstract form ensures that MMS interfaces defined for different programming languages have the same structure.

MMSI provides not only access to MMS services - both confirmed and unconfirmed - but offers a number of mechanisms to support this interaction with the MMS service provider. These are:

- Asynchronous call and event management: Requests can be issued asynchronously. This mechanism allows a function call to return immediately after the request has been queued for processing. A call may be issued some time later by the application to determine if the requested operation has been completed.
- Pairing of MMS request and response messages: When a confirmed service primitive has been issued, the interface is able to match the response with the original request and thereby to notify the application that the service request has been completed. This is the role of the so-called confirmed service provider of MMS, as shown in Figure 7.1.
- High level services: Some MMSI high level function calls execute a sequence of MMS services to achieve complex services such as downloading or uploading a domain from or into a file. The High Level Service Provider is the entity within MMSI responsible for the provision of these services. See Figure 7.1.
- Maintaining local objects as local types, local variables or local semaphores: MMSI provides facilities to create these objects with a local mapping in the user space memory, to access them when a service primitive is applied (Read/Write/Get attributes) and to delete them when they are no longer needed.
- Responder service: The responder service included in the interface can automatically recognise an incoming service request dealing with local objects, execute the expected action (Read/Write/Get Attributes) on these

objects and send the response to the peer MMS without the user application being involved in the processing.

- Memory and buffer management: MMSI provides support functions to dynamically allocate and de-allocate Data Control Blocks (DCBs) used as input and output parameters.
- Resource management: MMSI can monitor resource usage and notify the user when they have cleared up (Note When Clear function).

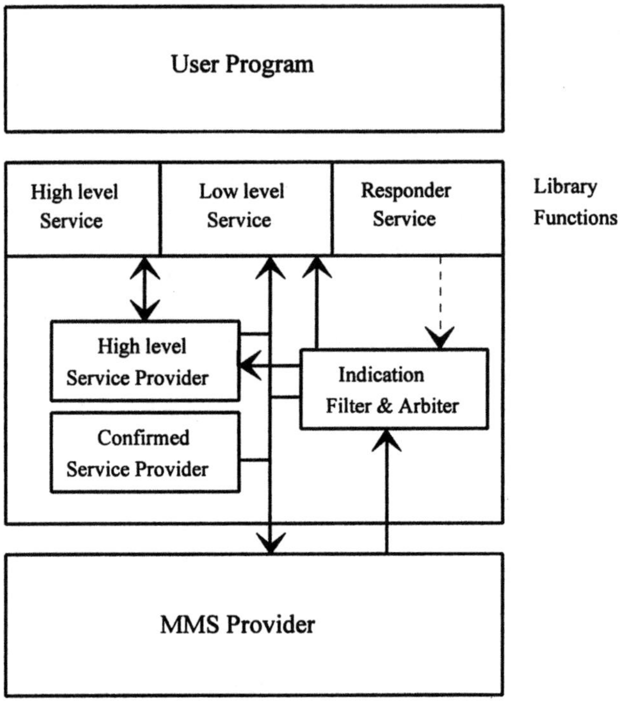

Figure 7.1: Architecture of the MMSI interface.

From a users point of view, MMSI is thus a set of functions that allow to allocate/deallocate data structures, used to access MMS services, to declare variables to the responder service or to create/delete local objects used in MMS services. More precisely, MMSI consists of:

- A set of functions that interface MMS services both as client and server.
- A set of support functions for event management, buffer management and error translation.
- A set of local functions used to build complex parameters to be passed through the interface or to manage local variables and semaphores maintained by the interface.
- A set of data structure definitions for input or output parameters.
- A set of error definitions.

7.4 Properties of MMSI

The MMSI interface is characterised by a number of properties that make it different from other interfaces to MMS. These properties are:

Operating system independence: All MAP APIs are operating system independent. They provide standard function calls to manage event handling and memory management which mask the operating system. The design of MMSI assumes a multi-tasking environment which supports inter-process communication and synchronisation.

Event-driven interface: To allow an application program to asynchronously interact with the network services, the interface service provider is modelled as a task distinct from the user program which issues function calls through the interface library. Network events are handled by the interface service provider concurrently with the user program. The interface provides a facility to notify the application when network events occur.

Synchronous and asynchronous modes: Function calls may be issued synchronously or asynchronously. In *synchronous mode* program control is returned when a response has been received for the service request. When performing a confirmed service, the user application program is suspended until the remote MMS peer has sent the response. In *asynchronous mode,* control is returned to the program immediately after the function call has been queued for execution. A parameter is used to associate an event with a previous asynchronous call. When the service request has come to completion, the interface service provider notifies the corresponding event to the user program. A wait function is provided to the user program to wait for event notifications. By making requests asynchronously, an application can greatly increase its efficiency if the time for the request and response to travel across the network is long.

High-level of functionality: The interface provides different levels of functionality to access MMS services (Figure 7.1). Low level services are used for sending confirmed and unconfirmed service requests, receiving indications, sending responses, sending and *pairing the request and response* messages. This corresponds to the level of service defined by the MMS standard. High level services can be performed, in one function call, grouping elementary functions to achieve complex services which require the exchange of a long sequence of MMS messages. Typical high level services would perform domain downloading, file transfer, or connection establishing with automatic retry until the remote device is ready. *A responder service* may be configured to maintain local objects such as variables or semaphores. When receiving a service request dealing with these local objects, the provider automatically accesses the values of these objects, executes the requested action and responds to the remote request without interacting with the user program.

Additional functionality: If a **reject indication** is caused by a remote protocol error and if it can be matched to an outstanding confirmed service request, the interface service provider is able to intercept this reject indication and *complete the pending request* with a reject protocol error. If the outstanding confirmed service request cannot be recognised, then the interface service provider aborts the connection. When an *abort indication* is received, the interface service provider *completes all outstanding requests* with an aborted connection error and then queues the abort indication on this association. This association is inactive but still exists for the user until he has processed the abort indication.

Object-oriented data representation: Certain MMSI functions require complex parameters as input or output. The MMSI interface provides a mechanism based on context-free functions to create or interpret MMSI objects corresponding to complex parameters. These objects are either built at the request of the user program and passed as input parameters, or returned by the interface as output parameters and can be interpreted by the user. They are not directly accessible by the user and are referenced by a unique identifier called a *handle.*

Buffer management: The interface provides a means to manage *Data Control Blocks (DCBs)* for use as input or output parameters. Support functions allow dynamic allocation and initialisation of DCBs with default values. When they are no longer needed a support function must be used to free them. The interface provides another facility to implicitly allocate output DCBs which are always large enough to store output data of variable size.

Strong error checking is performed on all input parameters by the interface to facilitate writing robust applications. Error conditions are reported with many levels of detail: error class, error code and vendor code to allow easy diagnostic and error recovery.

7.5 Classes of MMSI functions

MMSI defines a large number of functions which can be grouped into three main classes.

1. **Context-sensitive functions** depend for their invocation on the current state of the Application Entity involved in the communication process. It is not possible to call a Read request before a connection is established and an Application Entity has been activated. They provide access to the MMS services and result in primitives sent to or received from the network (PDUs). Context-sensitive functions can be invoked synchronously or asynchronously.
2. **Context-free functions:** They can be invoked at any time and do not depend on the status of the current Application Entity involved in the communication process. They are used to create and manipulate API objects that are subsequently used as input parameters in context-sensitive functions or to interpret objects previously returned as output parameters in context-sensitive functions. They do not result in messages sent to or received from the network. Context-free functions are always called synchronously.

3. **Support functions:** These are local functions common to all APIs. They cover four areas: Event management, DCB management, memory management and error translation.

Appendix 2 lists the context-sensitive functions defined in MMSI. Sections 7.7 to 7.9 describe the three function classes in more detail.

7.6 Naming rules for MMSI functions

The function naming rules are common to all the Application Program Interfaces specified by MAP. They intend to make the API easy to use:

- To distinguish which service the function is associated with MMS, FTAM, and ACSE.
- To meaningfully describe what the function actually does.
- To minimise the probability of collisions with identical names in other libraries.

The following conventions are applied:

- Function names are unique within the first 6 characters.
- No more than three words should be used.
- All words except the last are abbreviated using only their first letter.
- The last word is never abbreviated but may be truncated as required by the compiler used.
- Function names are lower case.

Function names are of the form shown in Figure 7.2.

ss_xxx[z]*( )	
ss	Service identifier: **mm** for MMSI **ft** for FTAMI **em** for Event Management ie: em_wait
_	Delimiter (language dependent)
xxx	Unique part of the function name up to 3 characters
[z]*	Part of function name which need not be unique

Figure 7.2: Form of Function names.

Table 7.1 provides examples of context-sensitive functions in MMSI

Function name	**Service**
mm_status()	Status
mm_sresponse()	Status response
mm_ireceive()	Indication receive
mm_idsequence()	Initiate Download Sequence
em_wait()	Wait for an event

Table 7.1: Examples of context-sensitive functions in MMSI

7.7 Context-sensitive functions

Context-sensitive functions implement MMS services and allow to manipulate MMS objects in the server VMD. They are grouped in functional subsets defined by MMS and called *Functional Units*:

- *Context Management :* Functions used for the establishment and release of associations between clients and servers.
- *VMD Support :* Functions giving access to information relative to the Virtual Manufacturing Devices (VMD) .
- *Domain Management:* Functions that allow for the transfer and manipulation of domains. These may contain data or program code.
- *Program Invocation Management:* Functions for the remote control of Program Invocations (PI) defined in remote VMDs.
- *File Management:* Functions that manage file transfer and management in the MMS VMD.
- *Variable Access:* Functions to access MMS Variables of various data types.
- *Operator Station Communication:* Functions that allow communication with an MMS Operator Station.
- *Event Management:* Functions allowing the creation and deletion of event conditions, actions and enrollments as well as the notification and acknowledgement of events.
- *Semaphore Management:* Functions that manage the MMS Semaphores used for co-ordinated use of resources and synchronisation between MMS clients.
- *Journal Management:* Functions to create and delete Journals and to read or write data in these Journals.

7.8 Support functions

Support functions provide general services to manage:

- Event reception associated to asynchronous calls when they come to completion.
- Allocation and deallocation of Data Control Blocks (DCBs) for input and output parameters.
- Deallocation of buffers dynamically allocated by the interface.
- Translation of errors into printable strings.

Table 7.2 lists the main support functions defined in MMSI.

Support Functions	
Function Name	**Description**
em_wait()	Wait For Event: Local Service to get the oldest queued return_event or wait for the next event corresponding to an asynchronous call that has come to completion.
mm_didcb()	Dynamic Initialise DCB: Local Service to allocate and initialise input or inout Data Control Block with default values.
mm_dfdcb()	Free DCBs dynamically allocated by a call to mm_didcb or implicitly by the interface in a context-sensitive function.
mm_nwcleared()	Note when cleared: Notify the user when a requested but up to now unavailable resource is available again.
mm_fdmemory() em_fdmemory()	Free Dynamic Memory: Local Service to free dynamic buffers allocated by the interface to store output data in context-free functions. (never DCBs)
mm_gperror() em_gperror()	Get Printable Error: Local Service to translate return_code and vendor_code into printable strings.

Table 7.2: MMSI support functions

7.9 Context-free functions

Context-free functions provide services to create MMSI objects subsequently used as input parameters in other functions or to interpret MMSI objects returned by other functions. The context-free functions may be grouped into many subsets:

- The general purpose functions for manipulating handles or lists of handles.
- The variable access support functions which allow to build or interpret parameters used in variable access functions (mm_read(), mm_write(), mm_ireport(), etc)

Tables 7.3 and 7.4 list the context-free functions used respectively for list management and variable access. Section 7.14 develops the notion of handle and describes in detail the operations that can be performed on them. Additional context-free functions are defined for journal management but are not listed in this book. The interested reader is referred to [MAP 91].

General Purpose Context-Free Functions	
Function name	**Description**
mm_mlist()	Make List: Local Service to add an element to a list of MMSI objects and create a new list if it does not exist.
mm_leextract()	List Element Extract: Local Service to remove an element from a list or delete the list if it is empty.
mm_dhandle()	Duplicate Handle: Local Service to duplicate a handle if logical copy or to create an identical copy of the object and return a new handle if a physical copy is required.
mm_fhandle()	Free Handle: Local Service to free a handle and delete the corresponding object if there is no logical copy of it.

Table 7.3: General-purpose context-free functions

Variable access context-free functions	
Function name	**Description**
mm_nmtype()	New Local Type: Local Service to define a new variable type.
mm_imtype()	Interpret MMS Type: Local Service to interpret a variable type handle and get the specification of type of a variable.
mm_nvspecification()	New Variable Specification: Local Service to define a variable access specification by providing the identifier of the variable.
mm_ivspecification()	Interpret Variable Specification: Local Service to interpret a variable specification handle and get the variable identifier.
mm_ndata()	New Data: Local Service to create an MMSI object that represents a data element of a variable value.
mm_idata()	Interpret Data: Local Service to interpret a data handle and get a data element of variable value.
mm_nlmapping()	New Local Mapping: Local Service to create an MMSI object that describes the local representation (type and local address) of a component of a structured variable.
mm_nnrepresentation()	New Network Representation: Local Service to create an MMSI object that represents the network type of a component of a structured variable.
mm_naaccess()	New Alternate Access: Local Service to create an MMSI object that specifies an alternate access to a component of a complex variable.
mm_iaaccess()	Interpret Alternate Access: Local Service to Interpret an Alternate Access handle and provide the component identifier of a complex variable that is to be accessed.
mm_nvassociation()	Define New Variable Association: Local Service to create an MMSI object that describes a variable with all its attributes (var spec, local mapping, network representation, data and data access error).
mm_ivassociation()	Interpret Variable Association: Local Service to interpret a variable association handle and get the information that describe a variable.
mm_vget()	Variable Get: Local Service to get a variable value from its local representation and add it to the MMSI variable association object. The mm_write() can implicitly call mm_vget().
mm_vput()	Variable Put: Local Service to put the variable data of a variable association object into its local representation. The mm_read() can implicitly call the mm_vput() function.

Table 7.4: Variable Management context-free functions

7.10 Parameters of MMSI functions

MMSI functions use a rich set of input and output parameters to exchange information with the MMSI service provider. Parameters of an MMSI function are either **exposed** in the parameter list, or embedded in a parameter called Data Control Block or **DCB**. MMSI distinguishes **input** parameters, **output** parameters, and the **return code** parameter that indicates whether the function completed successfully or not.

The general form of an MMSI function call is:

```
Return_Code = Function_Name(exposed_input_parameters,
                            input_DCB,
                            inout_DCB,
                            exposed_output_parameters)
```

The list of parameters vary according to the function called and may not include all the above fields.

MMSI distinguishes three types of parameters which may be passed as arguments in a function call either as input parameters or as output parameters:

- The exposed parameters.
- The Data Control Block parameters or DCB.
- The handle parameters.

7.10.1 Exposed parameters

Exposed parameters are parameters explicitly passed in the argument list of a function call. They are simple and mandatory parameters, generally common to many functions. They are obvious to the user and easy to set or get. For example, the connection identifier and the return event name provided in all context-sensitive functions are exposed input parameters, while the application entity label is an exposed output parameter for the Application Entity Activation function.

7.10.2 DCB parameters

The concept of Data Control Block parameter called DCB is extensively used in MMSI. DCBs are C data structures used to exchange input and output parameters that are specific to each function.

The DCB concept is intended to give a uniform view of the interface to the user by isolating function-specific parameters, by embedding complex parameters and by masking optional or conditional parameters which are not required in all applications. A second objective is to facilitate further extensions for companion standards that would add new optional parameters in a function without changing the interface.

In general, an MMSI function call uses two DCB parameters: an **input_dcb** for input-only parameters, and an **inout_dcb** for output parameters or for parameters that are used for both input and output.

Input and inout DCBs are not used in the same way by the interface.

Input DCBs may be immediately reused once the call has returned. When it must carry a large amount of data it is likely that the input DCB only contains a pointer to the actual data stored in the user memory space. The interface service provider consumes the input data before returning the control to the user program.

Output DCBs must not be reused before the completion of the function. Output data returned by a function in an inout DCB are only valid after the event indicating the completion of the function has occurred. Output data of variable length are stored in a variable length section of the inout DCB. A mechanism for dynamic allocation is provided by the interface to allocate implicitly an inout DCB always large enough to contain output data of variable length.

7.10.3 Handles

MMS lists of variables, MMS types, or MMS data may be considered as complex parameters which are not easy to directly pass to or return from the interface. These complex parameters may be considered as objects internal to the interface and referenced by an external identifier called "handle" which can easily be manipulated by the user at the interface.

As input, these objects must be built by the user with context-free functions and the corresponding handles can be passed as parameters in the input DCB. Then the handles are interpreted by the interface to access the objects and process the request.

As output, the interface builds the objects corresponding to the received data and returns the handles associated to these objects in the output DCB. The user has to interpret these handles using context-free functions to access the data contained in the objects.

A handle to an MMS object logically points to an MMSI object but it is not a pointer in the usual sense of a memory address.

7.10.4 Parameter order

To provide a consistent ordering for parameters listed in a function call, the following rules always apply [MAP 91]:

- Input parameters precede output parameters.
- Exposed input parameters precede the input DCB.
- If a parameter is used both as input and output parameter it is included in the input/output DCB.
- Exposed output parameters follow the input/output DCB.

7.10.5 Return code

The return code parameter reflects whether the issued function was successful or failed. It reflects a simple diagnostic information that can be expressed by an integer value. A return code value other than SUCCESS indicates that the call failed.

When a function is issued **synchronously**, the return code value reflects the final success or failure of the request.

When a function is issued **asynchronously**, the return code reflects whether the request has been successfully queued for processing. The actual return code is also included in the inout DCB of every function and is only valid after the user receives the event that notes the completion of the function by calling em_wait().

7.11 Synchronous and asynchronous modes

The MMS Application Programming Interface allows both synchronous and asynchronous calls and provides an event management function to wait for events that indicate the completion of asynchronous calls. It enables the user to have multiple asynchronous function calls outstanding at the same time and to manage in real-time the completion of these service requests. The general rule to observe is, all confirmed service calls expected to have a long delay should be executed in asynchronous mode.

7.11.1 Return event

An exposed input parameter called return_event_name specifies if the function call is issued in synchronous or asynchronous mode. If its value is SYNCHRONOUS the function call is issued synchronously otherwise the function is executed asynchronously.

In asynchronous mode, the return_event_name parameter is a unique event number which is assigned by the user to an asynchronous call. Control is returned to the program immediately after the request has been queued for execution. If the response to the service request has arrived, the em_wait() function returns the corresponding return event and the return_event_name can be reused for another asynchronous call.

7.11.2 Synchronous call

In a synchronous call, the return event name has the value SYNCHRONOUS. Control is returned to the calling program when the function is executed. The return code indicates whether the function call was successful and if output data are valid.

Figure 7.2 illustrates the notion of a synchronous call with the invocation of the MMS Status service. The MMSI function corresponding to this service is mm_status(). The parameters of this function are the connection identifier (cnx_id), the mode (SYNCHRONOUS), an input DCB (in_status_dcb) and an inout DCB (inout_stat_dcb), both specific to the Status service.

Example:

```
/* Issues a synchronous status service call which is a confirmed
service and waits until the response arrives */

return_code = mm_status(cnx_id,
                        SYNCHRONOUS,
                        &in_status_dcb,
                        &inout_stat_dcb);

/* When the function succeeds the status response is
available in the output data contained in the inout_stat_dcb */

if (return_code == SUCCESS)
        print_result(STATUS_SERVICE,inout_stat_dcb);
```

Figure 7.2: Example of a synchronous call in MMSI

7.11.3 Asynchronous call

In an asynchronous call, control is returned to the calling program immediately after the request has been passed to the MMS service provider. The application is then free to perform other tasks. The return code indicates whether the request has been successfully queued for processing or whether an error has been detected.

The return_event_name passed as input parameter must have a value other than SYNCHRONOUS which identifies the function call in a unique way. When the service request has come to a full resolution either because of the actual service completion or because of a non recoverable error, output data are updated in the inout DCB, the return code is returned and an event which matches the return_event_name parameter is notified by the interface service provider.

The em_wait() function allows to wait with a timeout for the occurrence of events noted when previous asynchronous calls come to completion. If events have already been notified, the em_wait() function returns immediately with the first incoming event. Otherwise, it suspends the calling program until an event occurs or a timeout occurs. The em_wait() function returns into the return_event output parameter the return_event_name of the first incoming event.

7.11.4 Example

The program shown in Figure 7.3 issues two asynchronous calls on two different connections, cnx_id1 and cnx_id2, to get the status of two different devices and handles the responses in the order they are received. A table of asynchronous call descriptors indexed by the return_event number is used to record the service and the output data corresponding to each request and access them when a return event is received.

```
/* declare a table to store the service and the output dcb associated
   to an asynchronous call indexed by the return event name */

s_async  async_call [ NB_MAX_ASYNC_CALL ];
int         nb_async_call_outstanding = 0;

/* store the service requested and set the address of the output dcb to NULL so
   that it should be dynamically allocated by the interface */

async_call [ RET_EVENT_1 ].out_dcb = NULL;
async_call [ RET_EVENT_1 ].service  = STATUS_SERVICE
async_call [ RET_EVENT_1 ].cnx        = cnx_id1;

/* call the first asynchronous request */

return_code = mm_status ( cnx_id1,
                          RET_EVENT_1,
                          status_in_dcb,
                          async_call [ RET_EVENT_1 ].out_dcb );

/* if error
   handle the return_code and the additional error diagnostic returned
   in the output DCB
  else
   increase the counter of outstanding asynchronous requests */

if ( return_code != SUCCESS )
{
  check_error ( return_code, &async_call [ RET_EVENT_1 ] );
  exit();
}
else nb_async_call_outstanding += 1;

async_call [ RET_EVENT_2 ].out_dcb = NULL;
async_call [ RET_EVENT_2 ].service  = IDENTIFY_SERVICE
async_call [ RET_EVENT_1 ].cnx        = cnx_id2;
return_code = mm_identify  ( cnx_id2,
                             RET_EVENT_2,
                             identify_in_dcb,
                             async_call [ RET_EVENT_2 ].out_dcb );
if ( return_code ! = SUCCESS )
{
  check_error ( return_code, &async_call [ RET_EVENT_2 ] );
  exit();
}
else nb_async_call_outstanding + = 1;
```

```
/* while all outstanding asynchronous call has not completed wait for an event or
   the timeout */
while ( nb_async_call_outstanding ! = 0 )
{
  return_code = em_wait ( &returned_event, timeout );
  switch ( return_code )
  {
   case SUCCESS:
     nb_async_call_outstanding - = 1;
     print_result ( &async_call [ return_event ] );
     break;

   case EME003_TIMEOUT;
     break;

   default:
     exit();
  }
}
```

Figure 7.3: Example of asynchronous calls in MMSI

7.12 DCB management

The input and inout DCB parameters of MMSI functions have a different structure and they must be managed in different ways. Two support functions mm_didcb() and mm_dfdcb() allow to manage allocation and de-allocation of DCBs and the interface provides a mechanism to implicitly allocate dynamic output DCBs always large enough to contain output data of variable length.

7.12.1 DCB types

Most of MMSI context-sensitive functions have one specific input DCB and one specific inout DCB that fulfils its special input and output requirements. Input and inout DCBs are defined by specific *DCB type declarations* in the header file mmsi_typ.h.

The DCB types are identified by a key which is defined in an enumerated value, as shown in the table below:

```
typedef enum Mm_dcb_type
{
    MMiAbort,
    MMoAbort,
    ...
    MMoIreceive,
    ...
    MMiRead,
    MMoRead,
    ...etc
}
```

DCB type constants are of the form:

MMiXxxx	for input DCB
MMoXxxx	for inout DCB

where **Xxxx** is the function name.

7.12.2 DCB allocation and deallocation

MMSI provides two support functions for DCB allocation and de-allocation. The mm_didcb() function is used to dynamically allocate memory and initialise fields with standard default values for a DCB. The type and size of the DCB are determined by the value of the dcb_type parameter. For inout DCBs the user must specify the size of the additional buffer with the additional_size parameter. This parameter should be large enough to fit incoming data of variable length. The address of the memory allocated for the DCB is returned in the dcb_pointer output parameter. The syntax for the mm_didcb function call is shown in figure 7.4.

```
return_code = mm_didcb (
                    Mm_dcb_type         dcb_type,
                    Uint32              additional_size,
                    Octet               **dcb_pointer
                    Api_rc              *result);
```

Figure 7.4: Syntax of a mm_didcb function call

The mm_dfdcb() function is used to dynamically free a DCB explicitly allocated by mm_didcb() or implicitly allocated by the interface when passing the address of a NULL pointer for the inout DCB. The syntax of this function is very similar to the one described above.

7.12.3 Using the input DCB

The main characteristics of an input DCB may be described in the following way:

- Input DCBs contain all specific input data required to perform a service request.
- Input DCBs may contain optional parameters and conditional parameters.
- Input DCBs may contain pointers to user data of variable length.
- All data contained or referred to by an input DCB are assumed to be handled or copied before the function call returns. That means that dynamic parameters stored in the stack can be passed in an input DCB.

The general structure of an input DCB is of the form:

Mask	Flags for optional input parameters
Specific Info Block	Input data specific to the function call

Figure 7.5 shows an input DCB for the MMS GetNameList service. The different fields of this structure are a mask used to indicate the presence or absence of optional parameters, and a number of parameters specific to this service.

Example:

```
typedef struct Mm_GNList_in_dcb
{
  Mask          mask;
#define         GNL_IN_DCB_CONTINUE_AFTER BIT(0)
  Mm_Context_id      context_id;
  Mm_Object_class    object_class;
  Mm_Object_scope    object_scope;
  Mm_Identifier      domain_name;
  Mm_Identifier      continue_after;
} Mm_GNList_in_dcb;
```

Figure 7.5: Example of use of the input DCB

An input DCB may be allocated in three different ways:

1. It may be both allocated and initialised *by the user*, either statically or dynamically by using the malloc() function of the C library, before calling the MMSI function.
2. It may be dynamically allocated and initialised with default values *by calling explicitly the* mm_didcb(), then modified to provide appropriate input data values in specific fields before calling the MMSI function.
3. It may be *implicitly* allocated and initialised with standard default values *by the interface* when the MMSI function is called with the inout_dcb parameter containing the address of a NULL pointer. In this case the MMS service is called with the standard default values. This implicit allocation of input DCBs must be reserved for some services which do not need to supply specific information.

An input DCB may be de-allocated or reused immediately after the function call returns. When it has been dynamically allocated by the user with the malloc() C function it must be freed by using the free() C function. When it has been dynamically allocated by the interface (implicitly or explicitly by calling the mm_didcb() function) it must be freed by using the mm_dfdcb() function.

7.12.4 Using inout DCBs

The main characteristics of an inout DCB may be described in the following way:

- Inout DCBs contain all specific output data returned by a service request: e.g. the data sent in a response of a confirmed service or the data received in a service indication.
- Inout DCBs can contain output data of variable size.
- Inout DCBs can contain optional parameters and conditional parameters.
- Output data contained in an inout DCB are valid:
  - either immediately after a synchronous call has returned.
  - or after the em_wait() has returned the corresponding return_event_name.

An inout DCB is made of three parts:

1. An item that contains the global size of the DCB.
2. A second part for fixed output parameters specific to the service. Its size depends on the type of the DCB.
3. A third part which is an additional buffer contiguous to the second part. This part is optional and only required if the service requested may return output data of variable length. It must be large enough to contain all variable length output data.

Table 7.5 shows the general structure of an inout DCB:

Size	Global size = fixed size + size of the additional buffer
Error Block	Block for error code, vendor code, additional error and additional error description. It may contain a pointer to the additional_description parameter when it is present.
Specific Info Block	Block for specific and fixed output parameters. It may contain pointers to data of variable length stored in the additional buffer.
Additional Buffer	Additional buffer for all output data of variable length. They are referred to by pointers contained in the first part.

Table 7.5: Structure of an inout DCB

An example of the use of an inout DCB is shown in Figure 7.7.

Example:

```
typedef struct Mm_Status_out_dcb
{
  Uint32                    size;
  Mm_Error_type             error_block;
  Mm_Status_block           status_info;
} Mm_Status_out_dcb;

typedef struct  Mm_Status_block
{
  Mask          mask;
#define         MM_SB_LOCAL_DETAIL                            BIT(0)
#define         MM_SB_COMPANION_STANDARD_DETAIL               BIT(1)
  Mm_Logical_status    vmd_logical_status;
  Mm_Physical_status   vmd_physical_status;
  Bit_string           local_detail;
  Mm_Comp_std_info     companion_standard_detail;
} Mm_Status_block;

typedef  struct  Bit_string
{
  Sint16       residue;        /* number of unused bits                 */
  Sint16       length;         /* length of the bitstring in octets     */
  Octet        *data;          /* pointer to the first octet            */
} Bit_string;
```

Figure 7.7: Use of an inout DCB

The status inout DCB must be allocated with an additional buffer large enough to contain data of variable length referred to by pointers in the error_block and the status_info parameters.

If the function succeeds the status_info parameter may contain the optional parameter local_detail that refers to a bit string by a pointer. In this case, the bit string of variable length is stored in the additional buffer of the inout DCB and the pointer data points to the first octet of the bitstring. This also applies to the companion_standard_detail and the additional_description parameters of the error block.

An inout DCB may be allocated and initialised in three ways:

1. **By the user:** He has to allocate a global memory space for the DCB plus its additional buffer, if required, and initialise the size in the first item of the DCB. The user is responsible for allocating a space large enough to contain all output data of variable length that may be returned by the function.
2. **Explicitly by the interface:** The user has to call explicitly the mm_didcb() function. In this case he has only to specify the type of DCB and the size of the additional buffer. The interface allocates the DCB plus its additional buffer. The user is responsible for specifying an additional size large enough to contain all output data of variable length and for freeing the DCB by calling to the mm_dfdcb() function when it is no longer needed.
3. **Implicitly by the interface:** When the user calls a context-sensitive function by passing the address of a NULL pointer as inout_dcb parameter, the interface implicitly allocates a dynamic inout DCB corresponding to the type required by the function and with a contiguous buffer always large enough to contain the output data of variable length. The user is responsible for freeing the DCB by a call to the mm_dfdcb() function when it is no longer needed.

7.13 MMSI main data types

MMSI functions use many predefined data types as parameters. These data types are specified in the C binding attachment of the MAP specification [MAP 91]. Data types which are *not prefixed* are types defined in the Context Management Attachment of the MAP 3.0 specification and are common to all MAP APIs. Those prefixed by mm_ are defined in the MMS Attachment of the MAP 3.0 specification and are specific to the MMS API.

They are declared in many header files that must be included when compiling an application:

mmsi_typ.h	Contains all basic data types plus the DCB type declarations.
mmsi_def.h	Contains all constants and default values for MMSI parameters.
mmsi_err.h	Contains return code, error code and vendor code values.

The following paragraphs do not discuss all MMSI data types and structures but concentrate on those necessary to understand MMSI parameters.

7.13.1 Basic data types

All the MMSI data types are derived from the following basic type definitions. The designers of MMSI precisely defined what signed and unsigned integers are, what a boolean (Bool) or octet string are. Figure 7.8 summarises the basic types defined in MMSI.

```
typedef unsigned char   Uint8;            /* 8-bit unsigned integer          */
                                          /* Range: 0 to (2**8) -1           */
typedef unsigned short  Uint16;           /* 16-bit unsigned integer         */
                                          /* Range: 0 to (2**16) -1          */
typedef unsigned long   Uint32;           /* 32-bit unsigned integer         */
                                          /* Range: 0 to (2**32) -1          */

typedef char            Sint8             /* 8-bit signed integer            */
                                          /* Range:-2**7 to (2** 7)-1        */
typedef short           Sint16;           /* 16-bit signed integer           */
                                          /* Range:-2**15 to (2**15)-1       */
typedef long            Sint32;           /* 32-bit signed integer           */
                                          /* Range:-2**32 to (2**32)-1       */

typedef Uint16          Mm_uint16;        /* Range: 0 to (2**16)-1           */
typedef Uint32          Mm_uint32;        /* Range: 0 to (2**32)-1           */

typedef Uint8           Bool;             /* Two-valued logical value:       */
                                          /* FALSE when all bits cleared*/
                                          /* TRUE otherwise                  */

typedef unsigned char   Octet;            /* 8-bit unformatted data          */
typedef Octet           *Octet_pointer;   /* Pointer to an Octet             */
```

Figure 7.8: Basic data types

7.13.2 Data types of exposed parameters

There are five different types of exposed parameters in MMSI functions. They are derived from the basic data types.

Mm_Vmd_id

```
typedef   Uint32          Mm_Vmd_id;        /* VMD Identifier */
```

A parameter of type Mm_Vmd_id is an identifier of a local VMD maintained by the interface which is unique regardless of the application process that manages it.

- It is returned as output parameter by the mm_vactivation() function that activates a local VMD and associates Application Entities.
- It is used as input parameter in many context-sensitive functions and some context-free functions in order to identify local objects attached to a local VMD.

Ae_label

```
typedef  Uint32          Ae_label;              /* AE_Invocation Identifier */
```

A parameter of type Ae_label is an identifier of a local Application Entity invocation which is unique regardless of the application process that owns it.

It is returned as output parameter by the mm_aeactivation() function that activates a local AE registered in the directory service before establishing a connection. It is used as input parameter to specify which AE invocation is requesting in the following functions:

- mm_connect() to establish a connection.
- mm_listen() to wait for incoming connection indication.
- mm_slisten() to stop listening.
- mm_aedeactivation() to deactivate an AE invocation.
- mm_aereset() to abort all connections established by an AE and to deactivate the AE.

Connection_id

```
typedef  Uint32          Connection_id;     /* Connection Identifier */
```

A parameter of type Connection_id is the identifier of a connection. It is unique regardless of the application process that owns it.

It is returned as output parameter by mm_connect() and mm_listen() when the connection confirmation or a connection indication is received. It is used as input parameter for most context-sensitive functions to specify on which established connection the MMS service request must be sent.

Local_event_name

```
typedef  Sint32          Local_event_name;            /* Return event name */
```

A parameter of type Local_event_name is the parameter that specifies if a function call is synchronous or asynchronous. If it has the value SYNCHRONOUS the function call is synchronous. Otherwise, it is a positive number that is associated to the pending asynchronous call until it has been completed and the corresponding return event has been returned by the em_wait function. Negative values are not permitted.

It is used as input parameter in all context-sensitive functions and returned as output parameter by the em_wait function.

Return_code

```
typedef  Sint32          Return_code;
```

Values of type Return_code are assigned to the function return code and in the return_code and vendor_code parameters of an inout DCB.

If it takes the value SUCCESS, this indicates either that the function has been correctly queued for processing in an asynchronous call or that the function succeeded in a synchronous call. If it takes another value it indicates an error either because of the failure to queue the request in an asynchronous call or because of a failure of the request itself.

7.13.3 DCB component data types

The following data types are other basic types used by MMSI in the DCB parameters.

Mask

```
typedef  Uint16          Mask;
```

A parameter of type Mask is a field of 16 logical flags assigned to optional parameters in a DCB or an embedded data structure and indicating, when a flag is set to 1, that the corresponding parameter value is valid.

Each optional parameter has a unique defined constant that allows to set or reset the corresponding bit in the mask parameter.

For example: optional parameters returned in an Initiate response are sent and received in a data structure of the following type, as shown in Figure 7.9.

```
typedef struct Mm_init_resp
{
 Mask        mask;

#define      MM_INIT_RESP_NEG_MAX_SEG_SIZE       BIT(0)

#define      MM_INIT_RESP_NEG_NESTING_LEVEL      BIT(1)

 Sint32                  neg_max_seg_size;
 Sint16                  neg_max_serv_calling;
 Sint16                  neg_max_serv_called;
 Sint8                   neg_nesting_level;
 Mm_init_resp_detail     init_resp_detail;
}Mm_init_resp;
```

Figure 7.9: Parameters of an Initiate response

If bit 0 or bit 1 are set to one this indicates that the neg_max_seg_size and the neg_nesting_level parameters have been sent by the remote user and are valid. Similarly, to send a response these bits can be selectively set in the mask as shown below.

```
mask = MM_INIT_RESP_NEG_MAX_SIZE |
       MM_INIT_RESP_NEG_NESTING_LEVEL;
```

Visible_string or Printable_string

```
typedef  char   *Visible_string;     /* visible character string */

typedef  char   *Printable_string;   /* printable character string*/
```

A parameter of type Visible_string or Printable_string is a pointer to a character string terminated by a NULL and should contain data that can be displayed or printed as ASCII characters.

Octet_string or Mm_Byte_string

```
typedef  struct   Octet_string        /* String of octets          */
{
  Uint32     length;                  /* Number of octets          */

  Octet_pointer pointer;              /* Pointer to an octet       */

} Octet_string;

typedef   Octet_string  Mm_Byte_string;
```

A parameter of type Octet_string or Mm_Byte_string is an octet string which can contain NULL values. It is defined by an Octet_pointer and the string length in octets.

Bit_string

```
typedef  struct  Bit_string           /* String of Bits            */
{
  Sint16       residue;               /* number of unused bits     */
  Sint32       length;                /* number of octets          */
  Octet        *data;                 /* first Octet contains MSB  */
} Bit_string;
```

A parameter of type Bit_string is a bit string octet aligned whose insignificant bits are at the end of the bit string. The data field points to the beginning of the bit string. The length field specifies the length in octets of the bit string and the residue field specifies the number of insignificant bits in the last octet.

7.14 MMSI objects and handles

MMSI objects and handles have been introduced to facilitate the handling of complex parameters and lists used in MMS services. They also make easier the handling of local objects as local variables, local semaphores or local journals that can be maintained by the interface.

7.14.1 Concept of handle

MMSI objects are objects *internal to the interface* which cannot be directly accessed by the user. They are *referenced by a unique identifier* called a **handle** which can easily be manipulated at the interface to build or interpret parameters of complex types or define local objects maintained by the interface like local variables, or local semaphores.

An MMSI handle may be compared to a file descriptor:

- A handle is not the MMSI object itself.
- A handle is not a pointer to an MMSI object in the sense of a memory address.
- A handle is a unique identifier that references an MMSI object managed by the interface.
- A handle has a positive integer value.

MMSI handles have been introduced to simplify the manipulation of complex parameters and local objects.

7.14.2 Handles and context-free functions

MMSI handles that refer to complex parameters can be used as input or as output parameters in context-sensitive functions. They are created or interpreted by calling context-free functions.

Complex parameters are built by the user with context-free functions and the corresponding handles are passed to the interface as input parameters of context-sensitive functions. Then the interface interprets and consumes these handles to get access to these parameters and perform the requested service.

In output of context-sensitive functions, complex parameters are internally built by the interface and the corresponding handles are returned in the output DCB. The user has to interpret these handles using context-free functions which access the values of these parameters and consume these handles.

Local objects as local variables, local semaphores or local journals that must be maintained by the interface can be declared at the interface by using context-free functions. Specific context-free functions allow to create, delete or get these local objects.

7.14.3 Creating a new handle

MMSI objects are created by context-free functions specific to each type of object. For example, mm_nvassociation() creates a new Variable Association object; mm_nvspecification() creates a new Variable Specification object.

The type of the object is implicitly known by the function call. The value of the attributes of the objects are specified in the exposed input parameter and the input DCB.

7.14.4 Making a list of handles

MMS parameters are often explicitly passed as lists - list of variable specifications, list of VMD attributes - or implicitly contain lists of components and may be recursively defined.

The Make List context-free function mm_mlist() allows to add an element to an existing list or to create a new empty list of a given element type and to append (or not) a first element. The result is a new handle to the list.

An element may be an MMSI object already existing and referenced by a handle or a new object to be created and inserted into the list. In this case it is described by a data structure and referenced by a pointer in the user space memory.

Two classes of MMSI list elements can be distinguished:

1. The class of "Handle" list elements. They have already been created by a mm_new_object() or the mm_mlist() functions and are referenced by a handle.
2. The class of "Data Structure" list elements. They are directly appended into a list by the Make List function from the user space memory. They are referenced by a pointer to a user data structure.

The elements on an MMSI list must have an homogeneous type: all the subsequent elements of the list must have the same type.

7.14.5 Extracting a handle or an object from a list

When context-sensitive functions return a list of objects as output parameters, the interface service returns a handle of type MMS_List. To get access to the subsequent elements, the List Element Extract function mm_leextract() allows to extract the first element of the list and returns a pointer to the list without this element.

7.14.6 Duplicating a Handle

The Duplicate Handle function allows to keep objects which must be reused in further function calls but whose handle would be consumed by the first function call. It offers two possibilities: the logical or the physical duplication.

Logical duplication provides a new handle that references the same object as the original handle. When one of these two handles is consumed the remaining handle still references the object. Figure 7.10 illustrates this type of duplication. A handle A pointing to an MMSI object is logically duplicated: A' points to the same object as A.

When using a logical copy of a handle, it is important to note that this handle references the same physical object and that if the user alters the object by a function call using one handle the other references the altered object. For example, if the user duplicates a list handle and extracts all the elements using the original list handle the logical copy handle references an empty list.

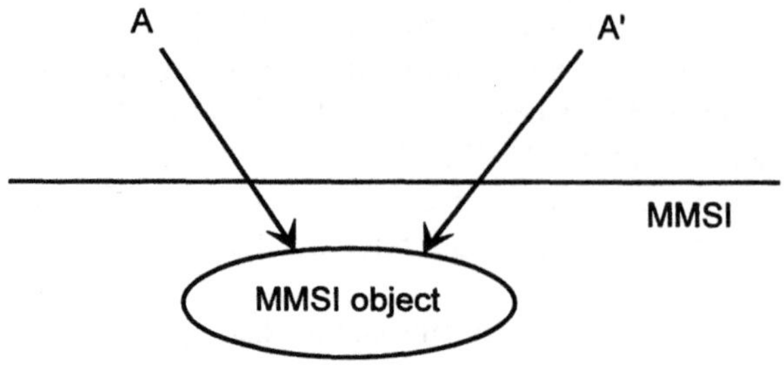

Figure 7.10: Logical copy of a handle

The physical duplication creates a physical copy of the object referenced by the original handle and returns a new handle to this new but identical object. If the object is a complex object all embedded objects are physically duplicated. Figure 7.11 illustrates physical duplication. A handle A is physically duplicated by creation of a copy of the object and a handle A'' to this new object.

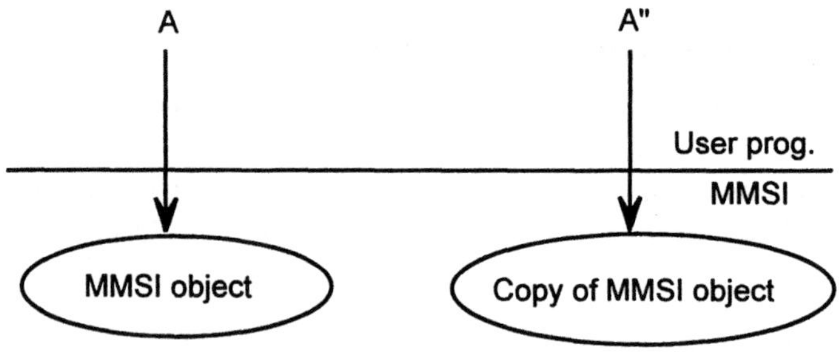

Figure 7.11: Physical copy of a handle

Note that logical copies reference the same object and physical copies reference separate but identical objects that may be used independently.

7.14.7 Freeing a handle

Context-free and some context-sensitive functions create objects referenced by handles. When handles are no longer needed, the Free Handle function, denoted by mm_fhandle(), allows to delete a handle associated to an object. If no other handles (logical copy) reference the object, the mm_fhandle() function deletes the object.

The Free Handle function is used when:

- The handle has not been consumed because the function failed.
- The user does not need to interpret the handle returned by a function.
- The user wants to delete a copy of a handle which is no longer needed.

7.15 Reading a variable

The purpose of this section is to illustrate the way a variable is read in MMSI, using all the concepts introduced here.

```
mm_nmtype();
mm_nvspecification();
mm_nvassociation();
mm_mlist();
mm_read();
```

Figure 7.12: reading a variable with MMSI (client side)

Let us examine what happens on the client side. First, a type object is created in the API with the mm_nmtype() function. This object describes the type of the variable that the user wants to read. The call returns a handle used later in the new variable association. Then, a variable specification is created in the API using the mm_nvspecification() function. The function call returns a handle to the variable specification object. This MMSI object indicates the name of the variable that will be accessed in the server. With the New Variable Association function, the user indicates to the interface where in the local memory space the value of the variable must be stored when the Read function call returns. The Make List service - mm_mlist() - is used because the read service expects a list as input. If the Make List service is used for the first time, it creates a list and includes the object referenced in the function call in this list. If the list already exists, the object is appended to the list. If multiple variables must be read, these four steps - new type, new variable specification, new variable association, make list - must be repeated for each variable in the list of variables to read. When the function call returns, the value of the variable(s) read are stored in the user space. After the function call, the interface objects are no longer usable because the interface has consumed the handles to these objects. If the user wants to reuse these objects in subsequent calls, he has to make copies of the handles to these objects, using for example the logical copy service of MMSI. Note that all these steps are independent of the way the call is made: synchronously or asynchronously.

7.16 Summary

MMSI is a standard application programming interface defined in the MAP 3.0 specification to access the MMS services. In this chapter, we have given a detailed introduction to this complex interface. The main advantages of a standard API are portability of applications and reduced training costs. MMSI is an interface providing high level services, strong error checking and an automatic responder. It is operating system independent and follows an object-oriented approach in the sense that interface objects can be created, deleted and linked to one another before calling interface functions. MMSI functions are divided into three groups: context-sensitive functions that correspond to the MMS services, context-free functions that are used to create and delete interface objects and support functions common to all services and used for event, memory and error management. Interface objects created by context-free functions are accessible by handles that are logical pointers to these objects. The chapter presented in more details the handling of synchronous or asynchronous calls, data control blocks and handles. Although complex, MMSI has one big advantage over all other interfaces: it is standardised in MAP which means that programs developed on one machine can be ported very quickly on another machine. Experiments made at the Swiss Federal Institute of Technology in Lausanne and at the University of Stuttgart confirm this claim of portability.

8. A DDE Interface to MMS

8.1 Introduction

There exists a very large base of standard application packages and tool kits in the Microsoft Windows environment, supporting the Dynamic Data Exchange (DDE) mechanism for communication between applications. These packages and tool kits allow users to produce software of high quality and low price.

This solution can either increase the features of a full manufacturing information system or provide solutions for more tailored applications. In general, these features are realised in a simple way.

The purpose of this chapter is to present a simple interface to MMS that is available in the Windows environment. The chapter describes an MMS/DDE server allowing direct access to manufacturing devices on the shop floor via MMS. The server described has been implemented within the ESPRIT CCE-CNMA project. MMS services provided in the MMS/DDE server cover Virtual Manufacturing Device services, Variable Management, Domain Management and Program Invocation Management services. An example is given at the end of the chapter to illustrate the use of this server from an Excel spreadsheet.

The next section Windows DDE basics, is taken from introductory material which can be found in [Microsoft 92a], [Microsoft 92b], and [Petzold 90].

8.2 Windows DDE basics

Dynamic data exchange (DDE) is the standard interprocess communications protocol under Windows 3.x. DDE allows communication between simultaneously running, co-operating Windows applications (e.g. Microsoft Excel, Lotus 1-2-3, Microsoft Word for Windows, Microsoft Access, Paradox, Microsoft Visual Basic, etc.). Figure 8.1 shows that DDE can be considered as communication channel between Windows applications. DDE provides the means to identify the peer with which an application wants to dialogue and to specify the resource of the peer the caller is interested to access. The exchanged data are transparent to DDE. An extension called NetDDE allows to exchange data between applications running on different PCs.

This section introduces the notion of DDE conversation, the way to initiate and terminate such conversations as well as the three kinds of links that can exist between DDE clients and servers.

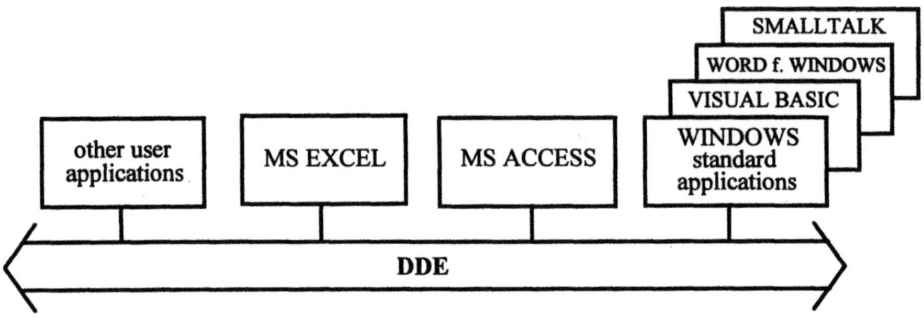

Figure 8.1: Dynamic Data Exchange under Windows 3.x

8.2.1 Client server principle

Two DDE applications enter into a DDE communication by exchanging data according to a fixed communications protocol. The roles of the two applications are fixed. The basis of the DDE protocol is Windows' own message system. The application which initiates a DDE communication and assumes control during the data exchange is called the **client,** the application which responds to the client's requests is the **server**. The client controls the channel, opens it, sends jobs and closes the channel. An application can support several DDE conversations acting as either server and/or client.

8.2.2 DDE conversation

The DDE communication between Windows applications takes place on a logical channel. The channel connects the client to the server. At the beginning of a DDE conversation, the client attempts to set up a link to the server and transfers the parameters **application** and **topic** with which a DDE conversation between client and server is defined. A DDE conversation is identified internally in Windows by the window handles of the client and server application. At the user interface of standard applications such as Microsoft Excel the DDE conversation is identified by a channel number.

8.2.3 The parameters application, topic and item

Three parameters identify a DDE conversation and the DDE data passed between client and server:

1. application
2. topic
3. item

The parameter *application* specifies the name of the server of the DDE connection in the Windows environment, e.g. for an MMS/DDE server 'ddemms'.

The parameter *topic* identifies an area for the data in the server to which the client requires access (e.g. the name of a local address information point that identifies all addressing information to set up an MMS connection, for an MMS/DDE server the name of the application association or with database applications the name of a database file). The topic depends on the particular server application.

If Microsoft's Network DDE [Brockschmidt 93] is used in order to connect to a remote DDE server on a different node, the destination node name must be supplied with the *application* parameter. For example:

1. The parameter *application* '\\my_PC\NDDE' would address an DDE server on the node my_PC from the DDE client application Microsoft EXCEL and,
2. The NDDE server at the remote node would retrieve through the parameter *topic* the application and topic parameters of the remote DDE server. The parameter *topic* would in this case address an entry in a configuration table at the remote node which contains the tuple 'application' and 'topic' for the remotely located DDE server.

Access to data defined in the server, which are to be transferred between client and server in a DDE communication is controlled by the parameters *application, topic* and *item*. With the parameter *item*, a data item defined in the server can be accessed during a DDE communication so that the values of the specified item can be transferred from the client to the server and vice-versa. The data must be transferred in a known clipboard format (e.g. CF_TEXT).

The syntax and the semantics of the DDE parameters *topic* and *item* are determined by the DDE server, i.e. a DDE client must know the way DDE is used by a server before it can communicate with it. Figure 8.2 shows the DDE parameters for *application, topic* and *item*.

	MMS/DDE server	MS Excel
application	"ddemms"	"excel"
topic	MMS connection	Excel table
item	MMS variable or service	field element

Figure 8.2: The DDE parameters *application, topic* and *item*

With the topic 'SYSTEM' and one of the following items general information about the DDE interface of a DDE server are available:

- SYSITEMS: List of items available for topic 'SYSTEM'.
- TOPICS: List of available topics.
- STATUS: Status of the DDE server (BUSY/READY).
- FORMAT: List of supported clipboard formats.
- HELP: General information about the DDE server, e.g., vendor name, version number, etc.
- ITEMS: List of all items supplied by the DDE server using other topics than 'SYSTEM'.

8.2.4 DDE messages

Windows applications can use DDE to exchange data and carry out remote commands through Windows messages. The DDE protocol defines exactly how DDE messages with their special parameters are exchanged between Windows applications. DDE does not define the meaning of the DDE parameters and the programming interface for a particular Windows application.

The following DDE messages can be exchanged in a DDE conversation. In the following, only the major sequences of DDE messages are described. A detailed description of the DDE protocol with all DDE messages can be found in the documentation of the Microsoft SDK - Software Development Kit [Microsoft92a, Microsoft92b].

WM_DDE_INITIATE:
With the WM_DDE_INITIATE message, a DDE conversation between client and server is initiated. The parameters *application* and *topic* are transferred to define the DDE conversation between the client and the server, as shown in Figure 8.3.

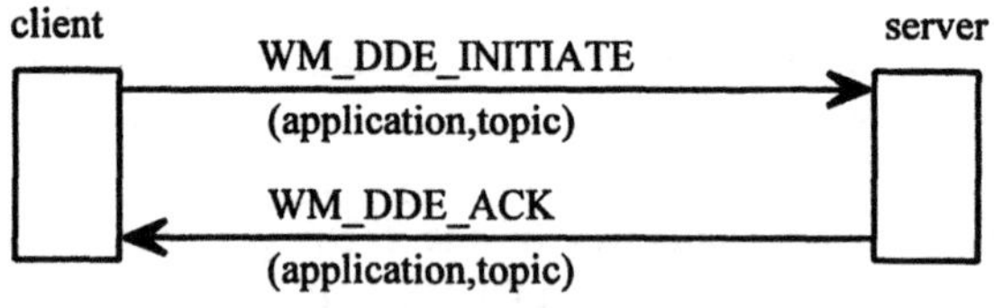

Figure 8.3: DDE message WM_DDE_INITIATE

WM_DDE_TERMINATE:
With the WM_DDE_TERMINATE message, an existing DDE conversation between a client and a server is terminated. As parameter, the identification of the DDE conversation is transferred. This relationship is shown in Figure 8.4.

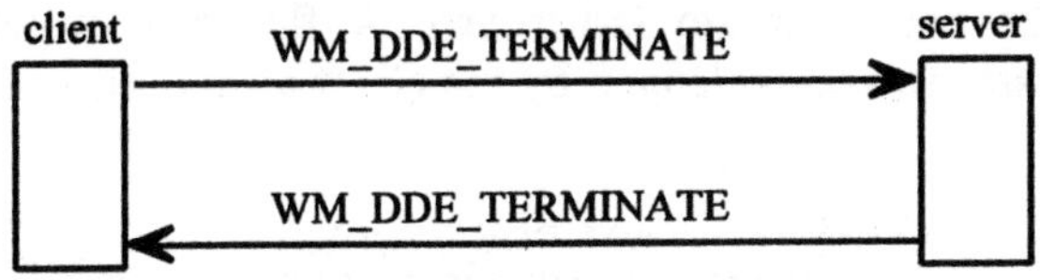

Figure 8.4: DDE message WM_DDE_TERMINATE

WM_DDE_ACK:
The WM_DDE_ACK message is sent during an existing DDE conversation as a response to certain messages. Examples of this message are shown in Figures 8.5 and 8.6. As parameters, apart from the identification of the DDE communication, the parameters *application* and *topic* are sent as response to a WM_DDE_INITIATE message, the parameters *status* and *command string* are sent as response to a WM_DDE_EXECUTE, and the parameters *status* and the name of the item as response to all other DDE messages. The *status* contains information about the type of acknowledgement (negative or positive WM_DDE_ACK) and, depending on the particular application, an error code.

WM_DDE_REQUEST:
With the WM_DDE_REQUEST message, the client requests the server to send the value of a specified item within an existing DDE conversation. As parameters, the identification of the DDE conversation, the name of the requested item and the clipboard format of the data to be sent are transferred. Examples of this message can be seen in Figures 8.5 and 8.6.

WM_DDE_DATA:
With the WM_DDE_DATA message, the server transfers the data requested with the WM_DDE_REQUEST or WM_DDE_ADVISE messages within an existing DDE conversation. As parameters, the identification of the DDE communication, the name of the requested item, the current value of the item and the clipboard format of the data sent are transferred. A flag also stipulates whether the transferred data are acknowledged by the client or not. This message is shown in Figure 8.5.

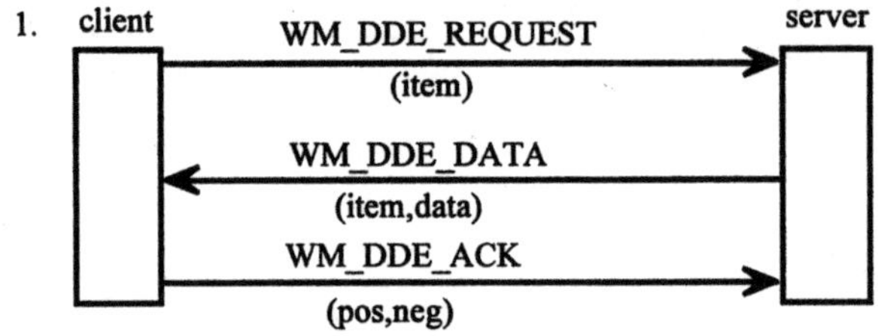

Figure 8.5: DDE message WM_DDE_REQUEST (pos,neg)

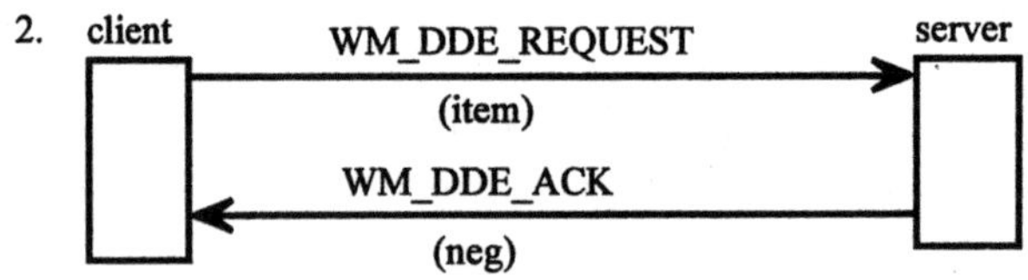

Figure 8.6: DDE message WM_DDE_REQUEST (neg)

WM_DDE_ADVISE:
With the WM_DDE_ADVISE message, the client instructs the server to transfer the value of a specified item or a signal if the value of the specified item changes during an existing DDE communication so that there is a permanent connection to the item. As parameters, the identification of the DDE conversation, the name of the connected item with additional information about values to be sent (e.g. the clipboard format and whether the transferred data are acknowledged by the client or not) are transferred. An example of this message can be seen in Figure 8.7.

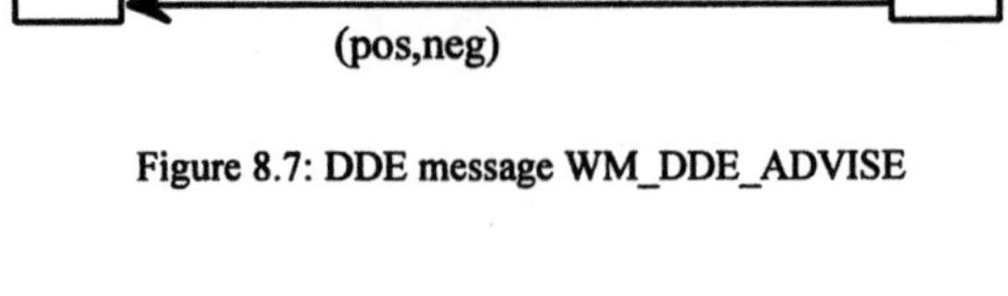

Figure 8.7: DDE message WM_DDE_ADVISE

WM_DDE_UNADVISE:
With the WM_DDE_UNADVISE message, the client terminates a permanent connection to a specified item within an existing DDE conversation. The DDE conversation is not terminated. As parameters, the identification of the DDE conversation and the name of the connected item are transferred. The client terminates connections to all items by using an empty string as item name.

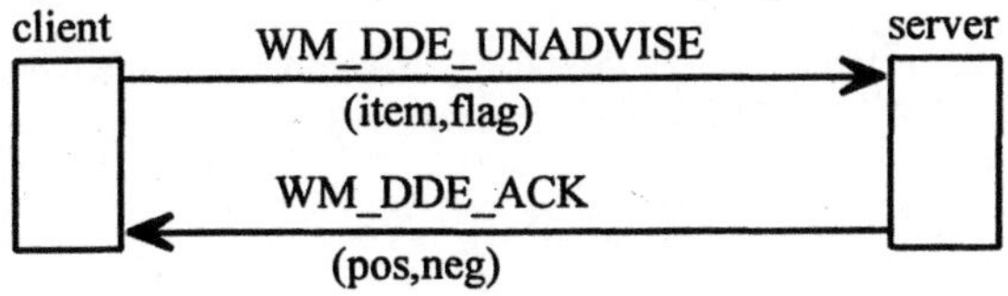

Figure 8.8 DDE message WM_DDE_UNADVISE

WM_DDE_POKE:
With the WM_DDE_POKE message, the client sends values of a specified item to the server during an existing DDE conversation. As parameters, the identification of the DDE conversation, the name of the current item, the value of the item and additional information about values to be sent (e.g. the clipboard format) are transferred. An example of this message can be seen in Figure 8.9.

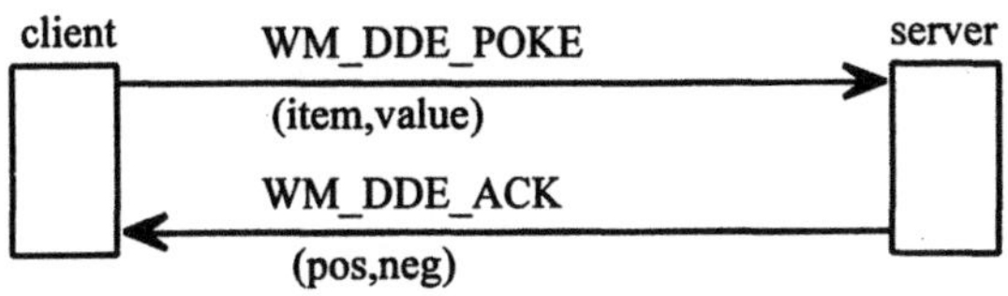

Figure 8.9: DDE message WM_DDE_POKE

WM_DDE_EXECUTE:
With the WM_DDE_EXECUTE message, the client sends a character string to the server during an existing DDE conversation which can trigger a series of commands in the server. As parameters, the identification of the DDE conversation and the character string with commands for the server are transferred. An example of this message can be seen in Figure 8.10.

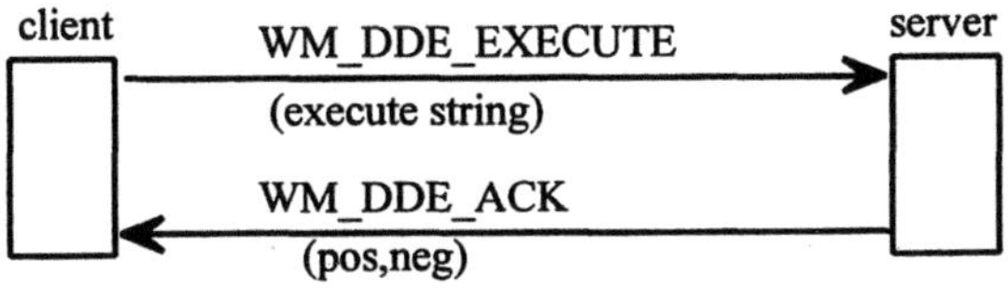

Figure 8.10: DDE message WM_DDE_EXECUTE

8.2.5 A typical DDE conversation

A typical DDE conversation proceeds as follows:

1. The DDE client initiates a DDE conversation using WM_DDE_INITIATE. The DDE server acknowledges the establishment of the DDE connection (WM_DDE_ACK) after receiving the message.
2. The applications exchange data according to the following rules:
   - If the server sends data (WM_DDE_DATA) to the client after the request (WM_DDE_REQUEST) this is known as a **cold link**.
   - The client can send data to the server without a request (WM_DDE_POKE).
   - On a **hot link**, the client instructs the server (WM_DDE_ADVISE) to send data (WM_DDE_DATA) whenever the data has changed.
   - On a **warm link**, the client instructs the server (WM_DDE_ADVISE) to send a signal (WM_DDE_DATA) whenever the data has changed. The data can then be requested by the client.
   - The client can instruct the server to execute a command (WM_DDE_EXECUTE).
3. The client or server terminates the DDE conversation using the message (WM_DDE_TERMINATE).

8.2.6 Cold link between client and server

With a cold link, the client uses the WM_DDE_REQUEST message to instruct the server to transfer the value of a specified item. If the server has the item available it sends the value of the item back in the message WM_DDE_DATA. The client can now confirm reception of the data with the WM_DDE_ACK message. Whether or not this acknowledgement is mandatory is determined by a flag in the WM_DDE_DATA message. If the server cannot meet the client's request it does not send the WM_DDE_DATA message but a negative WM_DDE_ACK.

The client's instructions with the WM_DDE_REQUEST message and the server's reactions with the WM_DDE_DATA and WM_DDE_ACK messages can be repeated as often as required for other data items until the client or server terminates the DDE connection with the WM_DDE_TERMINATE message.

Figure 8.11 illustrates these two scenarios.

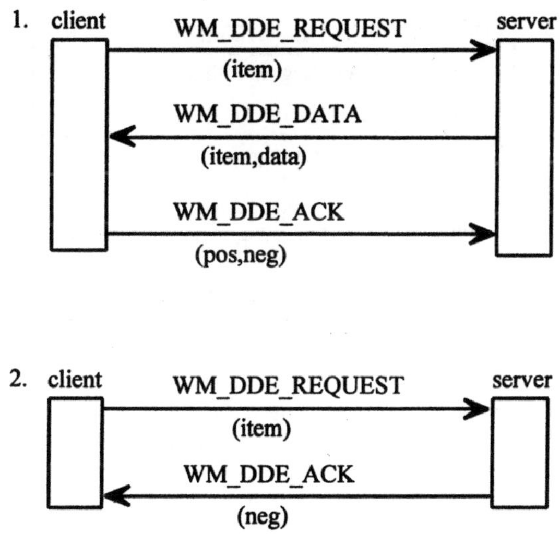

Figure 8.11: Cold link between client and server

8.2.7 Hot link between client and server

With a hot link between client and server, as shown in Figure 8.12, after establishment of the DDE conversation, the client requests a particular item with the WM_DDE_ADVISE message. If the server reacts with a positive WM_DDE_ACK, it confirms to the client that it will inform him if the value of the specified item changes. This information is conveyed by sending a WM_DDE_DATA message without this being specifically requested by the client. The client can now confirm reception of the data by sending the WM_DDE_ACK message to the server. Whether or not this acknowledgement is mandatory is determined in a flag in the WM_DDE_DATA message.

The data exchange is terminated by the client using the message WM_DDE_UNADVISE, which the server must acknowledge with a WM_DDE_ACK message. The DDE connection is terminated with a WM_DDE_TERMINATE.

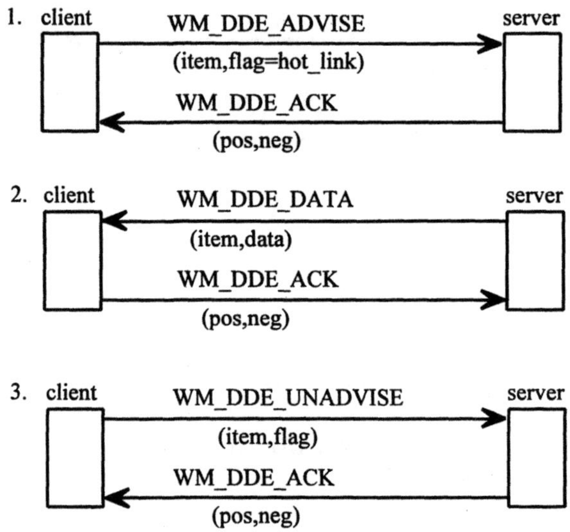

Figure 8.12: Hot link between client and server

8.2.8 Warm link between client and server

The warm link is a combination between a cold link and a hot link, in which, after establishment of the DDE conversation, the client requests a particular item with the WM_DDE_ADVISE message as shown in Figure. 8.13. In this case, the message is connected to a flag which stipulates that the server should only send a signal if the data changes. This signal consists of the WM_DDE_DATA message with the value NULL. The client, aware of the change, can now request the value of the item at a later point in time with a WM_DDE_REQUEST. The warm link mechanism is useful for a busy client.

The data exchange is terminated by the client with the WM_DDE_UNADVISE message, which the server must acknowledge with a WM_DDE_ACK. The DDE connection is terminated by a WM_DDE_TERMINATE.

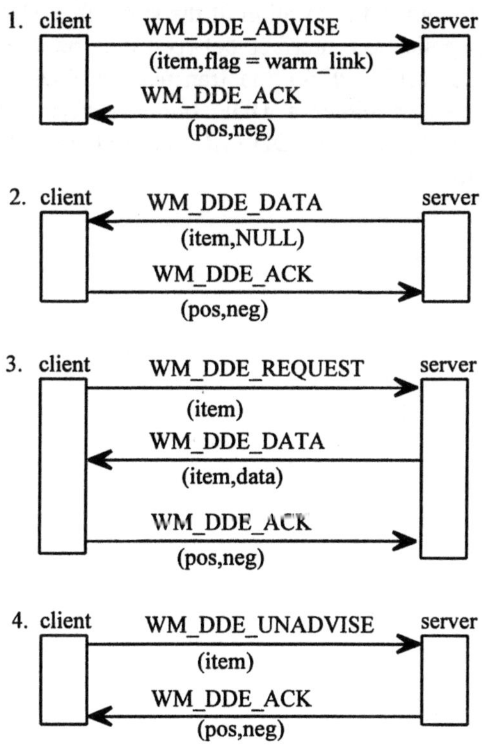

Figure 8.13: Warm link between client and server

8.2.9 DDE application interface

There exist a number of software development kits for various programming languages in the Microsoft Windows environment which can be used for programming and use of DDE, for example:

- Microsoft Windows Software Development Kit 3.1 [Microsoft92a, Microsoft92b]
- Microsoft Visual Basic [Microsoft93a] or
- Smalltalk/Win [Digitalk92].

The DDE application interfaces are specified in a language-independent, abstract form in terms of syntax and semantics of the DDE parameters *application*, *topic* and *item* of the corresponding DDE messages. Furthermore, it is described how the different DDE link mechanisms are used to access MMS objects. Figure 8.14 shows the location of the supported MMS/DDE services and its corresponding application interface between user interface of the DDE client application and the DDE-API for the transport of DDE messages. The DDE "transport" API may be either a DDE library offering a higher-level DDE procedural interface - for example DDEML - or the blank DDE message interface of Microsoft Windows.

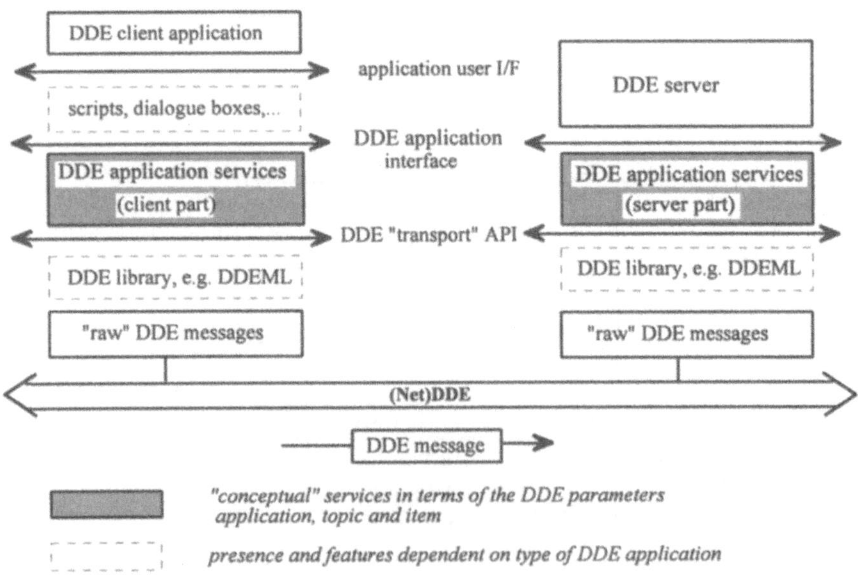

Figure 8.14: DDE-based Interface Hierarchy

The user interface that is offered to build these DDE messages and to control the DDE link mechanisms is dependent on the type of DDE client application. It may range from problem-oriented interactive dialogue boxes to script languages or 'C' function calls.

8.3 The MMS/DDE application interface

Within this section we provide an example of an MMS/DDE server that was implemented in the ESPRIT CCE-CNMA project [CCE 95]. The most useful services of MMS are supported in this implementation for the MMS/DDE server.

8.3.1 Overview

The MMS/DDE server allows direct access to manufacturing devices on the shop floor via MMS. MMS services provided through the MMS/DDE server are:

- Virtual Manufacturing Device services
- Variable Management services
- Domain Management services
- Program Invocation Management services

The following sections describe an abstract service interface for the MMS/DDE server which is primarily concerned with how the DDE protocol used to operate the MMS/DDE server.

The MMS/DDE server is always an MMS client. The service interface specifies primarily the semantics and the syntax of all DDE messages and their parameters which are of significance for the DDE protocol and the MMS/DDE server.

The DDE-based interface to MMS is specified in an abstract form via DDE messages. The DDE messages starting with "WM_DDE_" are placeholders for particular software development kits and/or other programming and user interfaces of standard applications which generate these messages in a different manner by appropriate procedure/function calls and call-back functions.

In the following, all parameters in the DDE messages described in *italic* letters are placeholders for variable values, all **bold** text described are fixed item strings. Optional parameters are enclosed in bold brackets **[ ]**.

All data or parameter values transferred in a WM DDE_REQUEST, WM_DDE_DATA or in a WM_DDE_POKE message are always ASCII strings. All data items, e.g. elements of an array, in this ASCII string are separated by <CR><LF> (i.e. <0xD><0xA>) and completed with <CR><LF><NULL> (i.e. <0xD><0xA><0x0>).

8.3.2 Establishing connections to MMS via the DDE parameter topic

To allow communication with MMS servers on the local area network an MMS connection needs to be established. The MMS/DDE server uses the DDE parameter *topic* for addressing purposes when setting up MMS connections. Through a mapping of the value of the parameter *topic,* all necessary addressing information is retrievable to set up an MMS connection together with the corresponding application association. This may be accomplished by either the MMS/DDE server, the MMS provider or other local means. Further below the parameter for this addressing purpose is specified as the name of the application association since the application association is established together with the MMS connection.

With the WM_DDE_INITIATE message, a DDE conversation between a DDE client and a DDE server is initiated. The initiation of a DDE conversation results in the establishment of an MMS connection. For each new *topic* a new MMS connection is established. For *topics* already in use (MMS connection already established) the same MMS connection is used.

WM_DDE_INITIATE (*application;topic*)

application	Name of an MMS/DDE server (e.g. **ddemms**)
topic	Name of an application association (e.g. **PLC1**)

WM_DDE_ACK (*channelNr;application;topic)*

application	Name of an MMS/DDE server (e.g. **ddemms**)
topic	Name of an application association (e.g. **PLC1**)

The *channelNr* is used here as an abstract specification aid in order to identify an established DDE conversation at the MMS/DDE application interface within the context of the initiating DDE application. As such, it is referred to in all subsequent DDE messages for this conversation. Its visibility in the application interface depends on the particular DDE client application.

8.3.3 Terminating a DDE conversation

With the WM_DDE_TERMINATE message, an existing DDE conversation between a DDE client and a DDE server is terminated. Both the DDE client and the DDE server may issue the WM_DDE_TERMINATE message. As parameter, the identification of the DDE conversation is transferred. The same WM_DDE_TERMINATE message is then sent back to the requesting DDE application to acknowledge the termination of the DDE conversation. As parameter, the identification of the DDE conversation is transferred.

The termination of a DDE conversation results in the closing of the associated MMS connection. If multiple DDE conversations were using one MMS connection, the termination of the MMS connection is performed when the last DDE conversation is terminated.

WM_DDE_TERMINATE (*channelNr*)

8.3.4 Status

The MMS service Status is used to request the state of a VMD. This consists of a logical and a physical status. The logical status indicates which services are supported in principle and the physical status indicates whether the real device is ready for operation.

WM_DDE_REQUEST (*channelNr*;**&status()**)

&status()	Fixed item representing the Status service.

WM_DDE_DATA (*channelNr*;**&status()**;*returnValue*)

&status()	Fixed item representing the Status service.
returnValue	ASCII String with logical and physical status separated by <CR><LF>.

logStatus Logical status of the VMD

STATE_CHANGES_ALLOWED	All MMS services permitted.
NO_STATE_CHANGES_ALLOWED	Only reading or interrogating MMS services permitted.
LIMITED_SERVICES_PERMITTED	Only Status and Identify permitted.
SUPPORTED_SERVICES_ALLOWED	All services permitted except of those not supported.

physStatus Physical status of the virtual device

OPERATIONAL	The real device is operational.
PARTIALLY_OPERATIONAL	The real device is partially operational.
NEEDS_COMMISSIONING	The real device is only operational on completion of local interventions.
INOPERABLE	The real device is not operational.

8.3.5 UnsolicitedStatus

The MMS service UnsolicitedStatus may be received from a VMD and forwarded to the DDE client. The status of a VMD consists of a logical and a physical status. The logical status indicates which services are supported in principle and the physical status indicates whether the real device is ready for operation. A DDE hot or warm link can be used to receive UnsolicitedStatus indications.

The request for a DDE hot link is transferred by the WM_DDE_ADVISE message with the *item* parameter set to the fixed item **&status()** and the warm link flag set to FALSE. The acknowledgement to the hot link request is returned in a WM_DDE_ACK message, whereas a status change already containing the new status value is transferred in the form of a WM_DDE_DATA message.

The hot/warm link to the status of a VMD is released by a WM_DDE_UN-ADVISE message with the *item* parameter set to the fixed item **&status()**.

These messages can be seen below.

- **Status Change Indicate Request**

WM_DDE_ADVISE (*channelNr*;**&status()**;*warmlinkFlag*)

&status()	Fixed item representing the service UnsolicitedStatus
warmlinkFlag	Boolean representing if a hot link (**FALSE**) or a warm link (**TRUE**) is to be established.

WM_DDE_ACK (*channelNr*;*adviseAck*);

adviseAck	Result of the requested hot/warm link to status of a VMD. If TRUE, the hot/warm link has been established, otherwise it has been rejected.

- **Status Change Indication**

WM_DDE_DATA (*channelNr*;**&status()**;*returnValue*)

&status()	Fixed item representing the Unsolicited-Status service.
returnValue	ASCII String with logical and physical status separated by <CR><LF>.

- **Status Change Release**

WM_DDE_UNADVISE (*channelNr;* ***&status()***)

&status()	Fixed item representing the UnsolicitedStatus service.

WM_DDE_ACK (*channelNr; adviseAck)*

adviseAck	Result of the hot/warm link release to the status of a VMD. If TRUE, the hot/warm link has been released, otherwise it has been rejected.

8.3.6 MMS variable services

Variable services allow to write to or read variables or to receive information reports (i.e. unsolicited variable data)

8.3.6.1 Variable types

The MMS/DDE server supports the reading and writing of variables for all MMS variable types and naming scopes (specific to VMD, domain and application association). The following types of variables are supported: boolean, integer, unsigned, float, bitstring, octetstring, visiblestring, time_of_day, time_and_date, structure and array.

8.3.6.2 Syntax of variable values

The values of variables transferred in a DDE conversation between an MMS/DDE server and a DDE Client are always ASCII-Strings in the CF_TEXT format. All input and output data are separated by <CR><LF> (<0xD><0xA>) and are completed with <CR><LF><NULL> (<0xD><0xA><0x0>). The following syntax is required for the permitted data types:

boolean: The boolean values "0" and "1" are permitted. A value not equal to 0 is interpreted as 1.

integer: Integers with 8, 16 and 32 bits with and without a sign are supported. Negative values are represented by the minus sign (e.g. -1). The MMS/DDE server automatically converts the type, i.e. when the user wants to write the value 0x0102 to an int8 variable, the variable receives the value 0x02 (0x as prefix indicates that the representation is hexadecimal).

unsigned: As for integers but without a sign.

floating: Strings representing real values according to the IEEE Single Precision Format are accepted. For example, "1.2", "3.45678" or "3.37E+05" are valid values.

bitstring: As bit strings, strings of zeros and ones are permitted. Bits exceeding the length of the bitstring are truncated. A valid input or output would be e.g. "0.1.0.1.0.0.0.1". A value not equal to 0 is interpreted as 1.

octetstring: A valid input or output would be e.g. "0c.0a.01.02". Otherwise the same applies as for bit strings

visiblestring: A visible string can be any string with ASCII characters, e.g. "Hello".

time_of_day: The format is "hh:mm:ss:msec" (e.g. "13:05:13:100" - 1 p.m. 5 minutes 13 seconds and 100 milliseconds). Seconds and milliseconds are optional to support the Excel date format.

time_and_date: This type is an extension of the time_of_day adding the date. The format is "dd.mm.yyyy mm:ss:msec" (e.g. "1.2.1992 13:05:13:100" - 1st. February 1992, 1 p.m. 5 minutes 13 seconds and 100 milliseconds). Seconds and milliseconds are optional to support the Excel date format.

structure: The value buffer of a structure consists of the components of the structure. The individual values are stored in the value buffer in the order in which they were defined and separated by <CR><LF>.

array: The value buffer of an array consists of the components of the array. The individual values are stored in the value buffer in the order in which they were defined and separated by <CR><LF>. (e.g. an array var[4] with the integer values 9. 7. 4. and 3: "<CR><LF>9<CR><LF>7<CR><LF>4<CR><LF>3<CR><LF><CR> <LF> <NULL>").

8.3.6.3 Access and alternate access

The MMS/DDE server supports only the reading and writing of a variable by name. The reading or writing of the whole variable:

```
struct {
  integer temp;
  boolean switch[2];
} sensor;
```

is made with the name **'sensor'** as item.

Since some Windows applications do not recognise structured data types (e.g. Intouch), the MMS/DDE server supports also the alternate access to variables, so that the components of a variable can be manipulated by alternate access. Alternate access to the example variable **'sensor'** is made by **'sensor.temp'**, **'sensor.switch[0]'** or **'sensor.switch[1]'** as item.

To access the structure components the **period** **"."** is used and to access the elements of an array the **square brackets open/close** **"["** and **"]"**. Upper and lower case letters are recognised and must be kept in the variable names.

Structures and arrays are supported up to **nesting level 2**. This allows structures containing arrays or arrays of structures to be read and written. Access to a variable with a nesting level higher than 2 is rejected.

8.3.6.4 Reading a variable

With the Read service, the client can read a simple or a complex variable.

WM_DDE_REQUEST (*channelNr*; *varName*)

varName	Name of the variable.

WM_DDE_DATA (*channelNr*;*varName*;*returnValue*)

varName	Name of the variable.
returnValue	ASCII String with value of a variable separated by <CR><LF>.

8.3.6.5 Writing a variable

With the Write service, the client can write to a simple or a complex variable.

WM_DDE_POKE (*channelNr*;*varName*;*VarValue*)

varName	Name of the variable.
VarValue	ASCII String with values of a variable separated by <CR><LF>.

WM_DDE_ACK (*channelNr*;*pokeAck*);

pokeAck	Result of the write operation. If TRUE, the write operation link has been done, otherwise it has been rejected.

8.3.6.6 Notification of variable value changes (Information Report)

MMS information reports are notified to the DDE client after a hot link or warm link has been established with the DDE message WM_DDE_ADVISE. Only value changes are notified. The required messages are shown below. The DDE hot link or warm link are used to receive MMS information reports.

The request for a DDE hot link is transferred by the WM_DDE_ADVISE message with the *item* parameter set to the variable name and the warm link flag set to FALSE. The acknowledgement to the hot link request is returned in a WM_DDE_ACK message, whereas a variable change already containing the new variable value is transferred in the form of a WM_DDE_DATA message.

The request for a DDE warm link is transferred by the WM_DDE_ADVISE message with the *item* parameter set to the variable name, and the warm link flag set to TRUE. The acknowledgement to the warm link request is then returned in a WM_DDE_ACK message, whereas the variable change is just indicated in the form of a WM_DDE_DATA message without accompanying data. In this case, the new variable value must be explicitly requested by the DDE client using the WM_DDE_REQUEST message.

The hot/warm link to the variable of a VMD is released by a WM_DDE_UNADVISE message with the *item* parameter set to the variable name.

- **Variable Value Change Indicate Request**

WM_DDE_ADVISE (*channelNr*;*varName*;*warmlinkFlag*);

varName	Name of the variable.
warmlinkFlag	Boolean representing if a hot link (**FALSE**) or a warm link (**TRUE**) is to be established.

WM_DDE_ACK (*channelNr*;*adviseAck*);

adviseAck	Result of the requested hot/warm link to a variable. If TRUE, the hot/warm link has been established, otherwise it has been rejected.

- **Variable Value Change Indication**

WM_DDE_DATA (*channelNr*;*varName*;*returnValue*)

varName	Name of the variable.
returnValue	ASCII String with value of a variable separated by <CR><LF>.

Note: The *returnValue* is only significant in case of a hot link. In case of a warm link, the new value of the status must be explicitly requested by a subsequent WM_DDE_REQUEST.

- **Variable Value Change Release**

WM_DDE_UNADVISE (*channelNr;varName*)

varName	Name of the variable.

WM_DDE_ACK (*channelNr; adviseAck)*

adviseAck	Result of the hot/warm link release to the variable. If TRUE, the hot/warm link has been released, otherwise it has been rejected.

8.3.7 MMS domain download

Domain services allow memory data or program data to be downloaded from a file to the VMD or upload domains from the VMD into files. Domain downloads are composed of three phases: initiation, transfer and termination of the download. MMS services allow the client to have direct control of the different phases of a download. The MMS/DDE interface for domain downloading is simplified and provides a high level interface as the following primitives show:

WM_DDE_REQUEST (*channelNr*; **&download(***domName*;*FileName***)**)

&download(...)	Fixed item representing the MMS service sequence Downloading
domName	Name of the domain to be loaded.
fileName	Name of the local file, in which the domain is stored.

WM_DDE_DATA (*channelNr*;**&download(***domName*;*fileName***)**,*returnValue*)

&download(...)	Fixed item representing the MMS service sequence downloading.
returnValue	Status of the domain on completion of the operation (Success=**FIN**).

With this approach, the client does not see the three phases that compose a download. All necessary sequencing of MMS services, including file access is automatically performed by the MMS/DDE server. The client specifies the name of the domain to load and the file containing the data to transfer. In response, the server indicates the result of this operation.

8.3.8 Further MMS services supported

The MMS/DDE server functions as described above are examples for the supported MMS services. A full description of the functions implemented in the ESPRIT CCE-CNMA project can be found in [CCE 95].

8.4 Operation of the MMS/DDE Server

8.4.1 Introduction

This section describes the operation of the MMS/DDE server. This server acts as a DDE server to service requests from Windows applications and provides the mapping of the DDE messages to an MMS provider. The MMS/DDE server always acts **only** as an MMS client for interfacing to MMS.

This chapter describes:

- The mapping of DDE messages to MMS services by the MMS/DDE server.
- The additional DDE server functionality for the continuous reading (polling) of MMS variables, for sending notifications of changes in variables and automatic sequencing of MMS services for domain transfers.

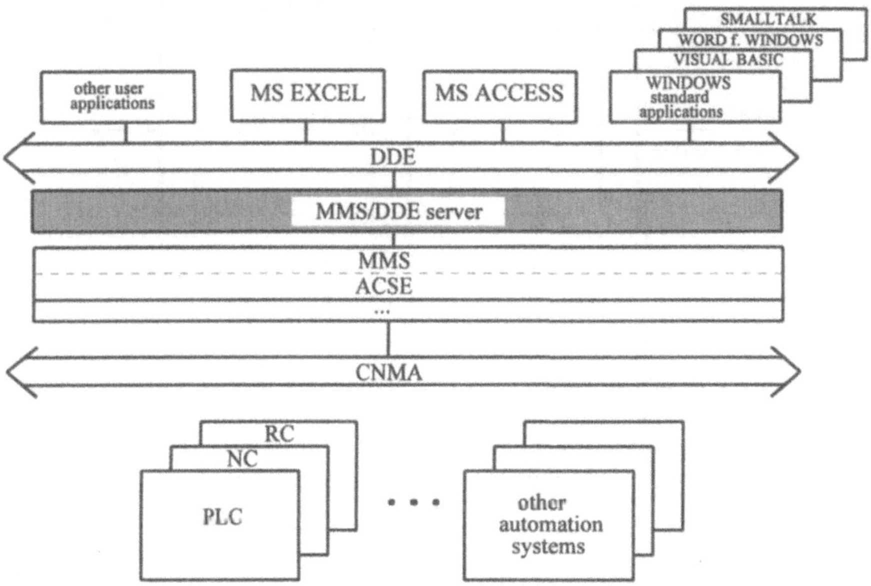

Figure 8.15: Functional Architecture of the MMS/DDE server

Figure 8.15 illustrates the positioning of the MMS/DDE server in a Windows environment. This server acts as an interface between Windows applications and the MMS provider. Communication with the server relies on the DDE communication mechanism while communication with manufacturing devices relies on the MAP or CNMA communication networks.

8.4.2 Initiating a DDE conversation

The initiation of a DDE conversation results in the establishment of an MMS connection. For each new *topic* a new MMS connection is established. If for a particular *topic,* an MMS connection is already established the same MMS connection is used. Figure 8.16 illustrates the establishment of an MMS connection using the WM_DDE_INITIATE message and the associated *application* and *topic* parameters.

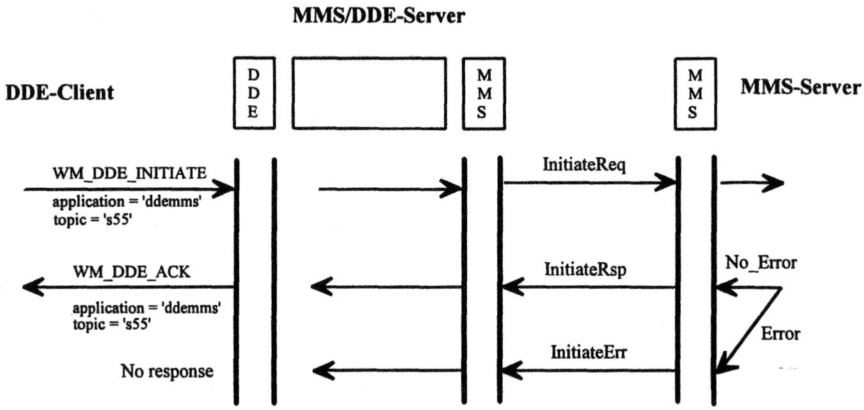

Figure 8.16: Initiating a DDE Conversation

In Figure 8.16 and subsequent figures, the first vertical double bar represents the DDE interface, the space between the first and second bar the MMS/DDE server while the space between the second and third double bars corresponds to the communication network with the MMS PDUs transferred between the client and the server.

8.4.3 Terminating a DDE conversation

The termination of a DDE conversation results in the closing of the associated MMS connection. If multiple DDE conversations were using one MMS connection, the termination of the MMS connection is performed when the last DDE conversation is terminated. Figure 8.17 illustrates the termination of a DDE conversation with the WM_DDE_TERMINATE message.

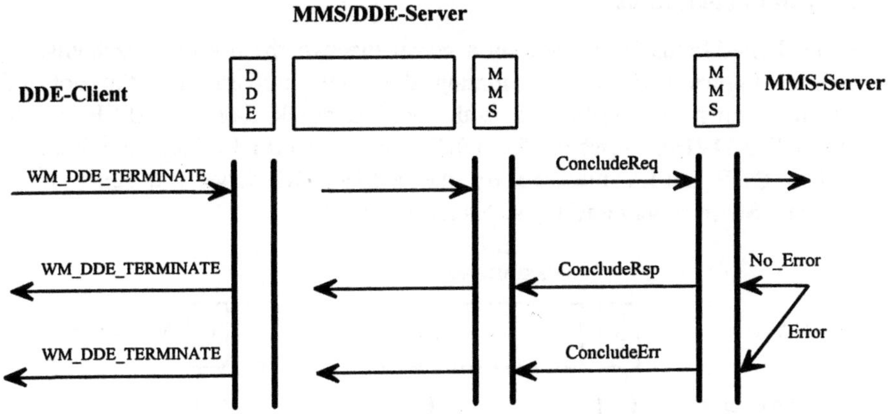

Figure 8.17: Terminating a DDE Conversation

8.4.4 Status

Figure 8.18 depicts the actions performed by the server when receiving a Status request in the WM_DDE_REQUEST message. The server translates this into a StatusRequest PDU and waits for a response. Two outcomes are possible: a StatusResponse results in a WM_DDE_DATA message with the physical and logical status's; a Status Error results in a WM_DDE_ACK message with a negative status.

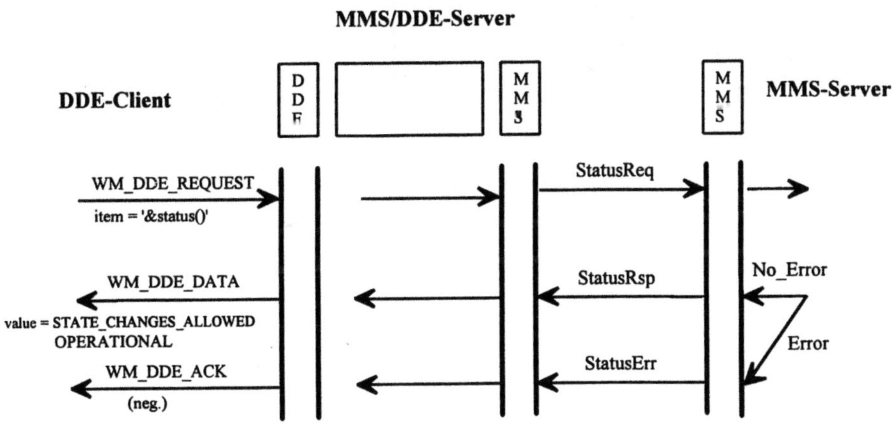

Figure 8.18: Mapping of the MMS Status Service

8.4.5 UnsolicitedStatus

An UnsolicitedStatus is reported by a server through the hot link mechanism as shown in Figure 8.19. After initiating the DDE conversation, the hot link mechanism must be established with the 'warmlink_flag = FALSE' in the WM_DDE_ADVISE message. The DDE client is notified of this condition via the WM_DDE_DATA message by the MMS/DDE server but only if an UnsolicitedStatus is sent **and** the status has changed.

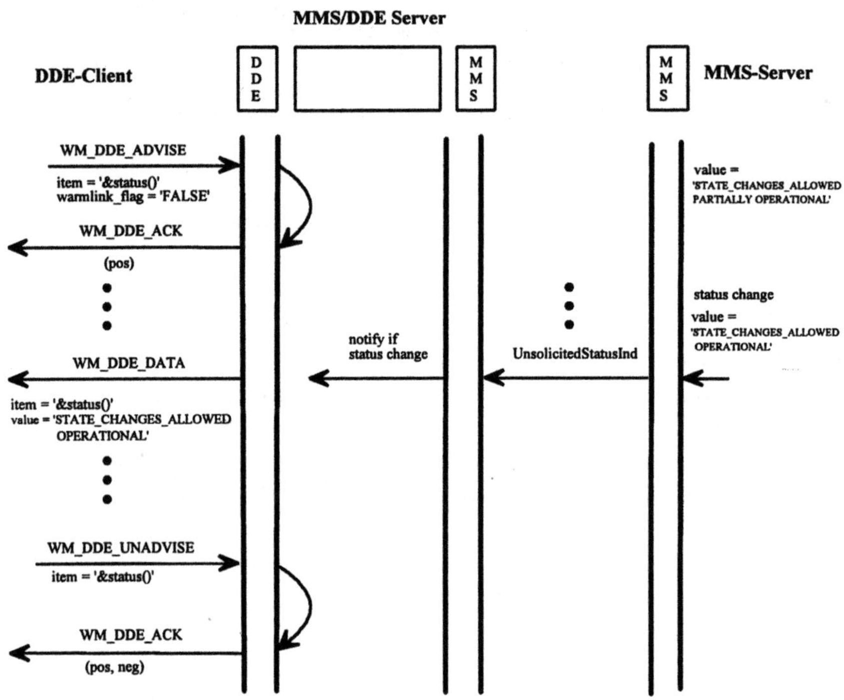

Figure 8.19: UnsolicitedStatus via hot link support

8.4.6 MMS variable services

The purpose of the MMS Variable Services are to write or read variables or to receive information reports (i.e. unsolicited variable data). The sequencing and mapping of the DDE messages to the MMS variable services are shown below. The MMS Variable services supported by the MMS/DDE server are shown in Table 8.1 below.

MMS service:	**Direct support via DDE message:**
Read	WM_DDE_REQUEST
Write	WM_DDE_POKE
InformationReport	WM_DDE_ADVISE
GetVarAccessAttr	WM_DDE_REQUEST

Table 8.1: Mapping of the MMS Variable services

The MMS/DDE server provides different means of reading variables:

- MMS variable read via cold link initiated by the DDE client for each read.
- Notification of variable value changes on receipt of MMS information reports from the VMD.
- Notification of variable value changes based on continuous read (polling) of the MMS/DDE server.

The supported features of MMS variables are:

- named variables
- structures
- arrays
- nesting level = 2
- alternate access
- name scopes for variables are VMD specific, domain specific and application association specific
- reading the type attributes of a variable

8.4.6.1 Reading the value of a variable on a cold link

The DDE client needs to issue for each variable a WM_DDE_REQUEST which is mapped by the MMS/DDE server to an MMS Read service. Depending on the result, the answer is returned to the client either as a WM_DDE_DATA message with the value of the variable coded in ASCII or as a WM_DDE_ACK in the case of an error. Figure 8.20 illustrates this scenario.

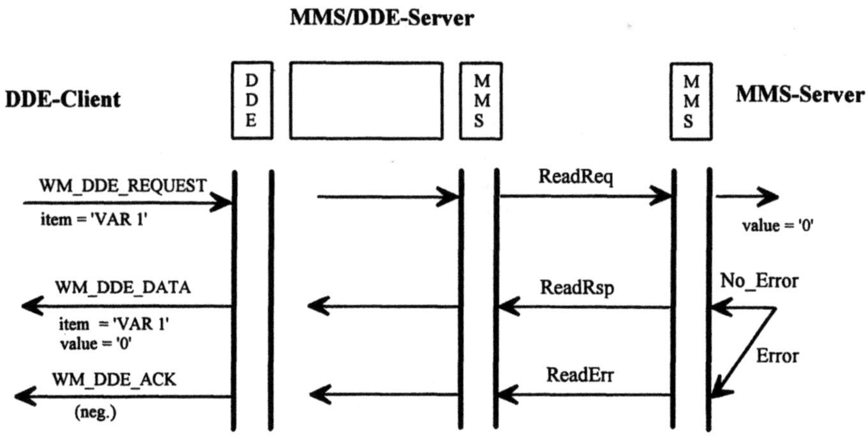

Figure 8.20: Reading the value of a variable via cold link

8.4.6.2 Reading the value of a variable via hotlink (Information Report)

If a hot link has been set up with the WM_DDE_ADVISE message by a DDE client to a variable, the MMS/DDE server reports with the DDE message WM_DDE_DATA all new variable values received by MMS information reports. Figure 8.21 illustrates the process.

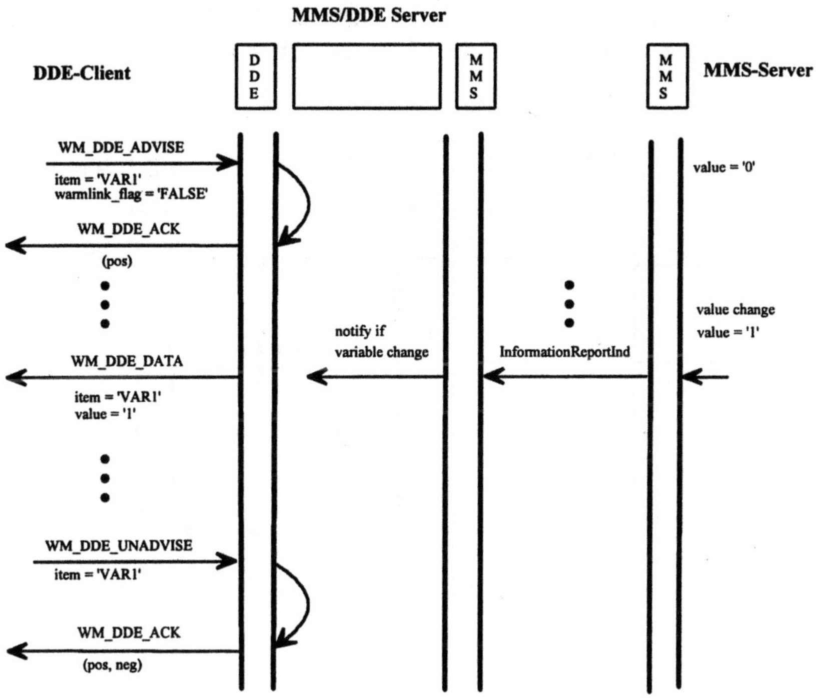

Figure 8.21: Reading the value of a variable via hot link (Information Report)

8.4.6.3 Reading the value of a variable via hotlink (Polling)

If the VMD is not able to issue information reports the MMS variables in question may be polled by the MMS/DDE server and reported to the DDE client through a hot link. Only value changes are reported. After initiating the DDE conversation, the hot link mechanism must be established with the 'warmlink_flag = FALSE' within the WM_DDE_ADVISE message. The polling cycle is passed with the parameter *topic* in the DDE message WM_DDE_INITIATE. Figure 8.22 illustrates the sequence of calls to and replies from the DDE server.

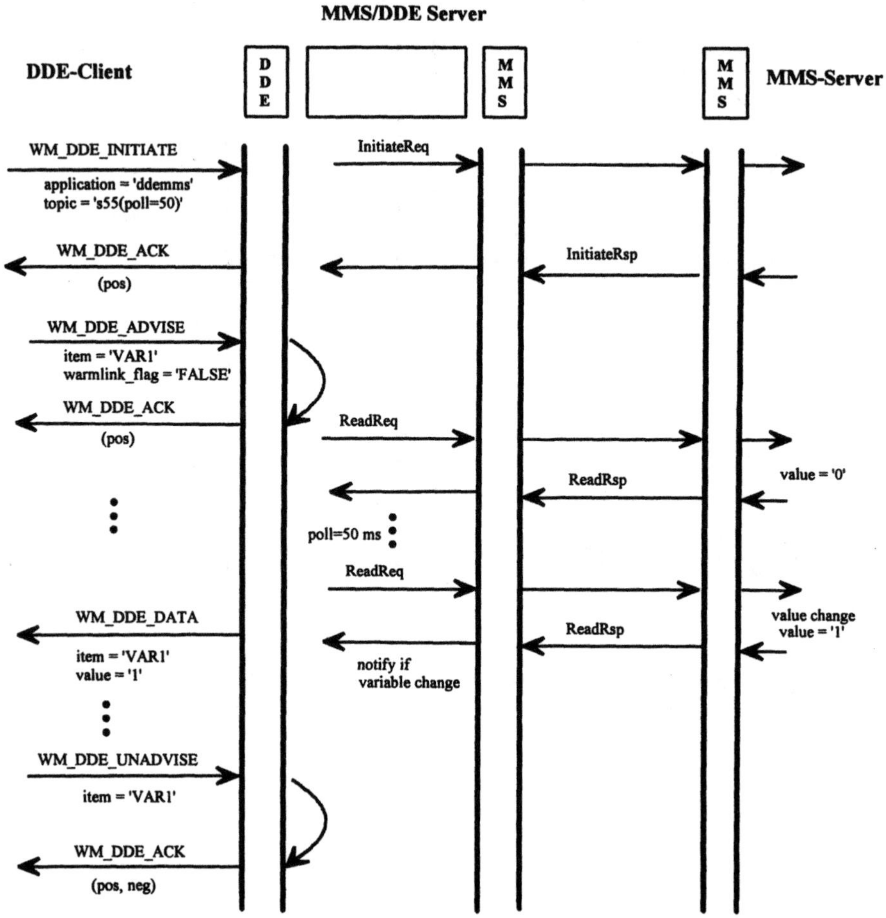

Figure 8.22: Reading the value of a variable via hot link (Polling)

8.4.6.4 Writing the value of a variable

The DDE client needs to issue for each variable to be written the DDE message WM_DDE_POKE which is mapped to the Write service. The sequencing is shown in Figure 8.23.

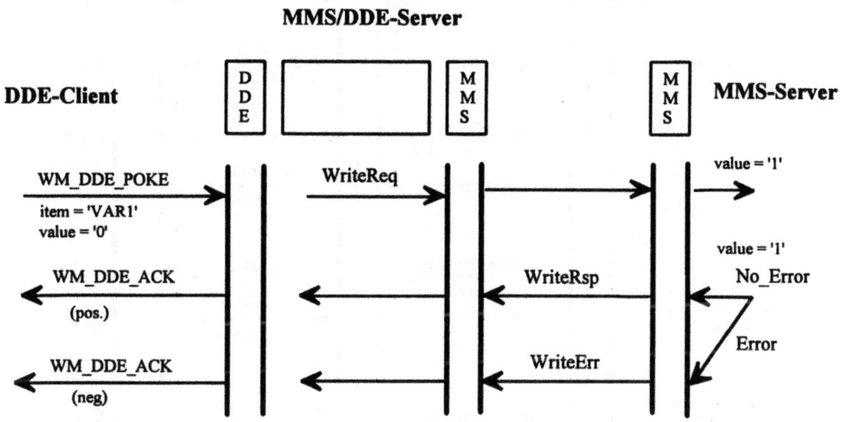

Figure 8.23: Writing the value of a variable

8.4.7 Downloading a domain

Downloading a domain from a local file is initiated by the DDE message WM_DDE_REQUEST. All sequencing of MMS services for downloading the domain is performed by the MMS/DDE server which provides to the user a high level interface hiding the details of the underlying MMS domain services. Figure 8.24 illustrates the operation of the MMS/DDE server for the download of a domain from the DDE client to the MMS server. The normal outcome indicates a successful download with the WM_DDE_DATA message. In case of error, at any point in the download procedure, a WM_DDE_ACK message is returned to the client (bottom part of the figure).

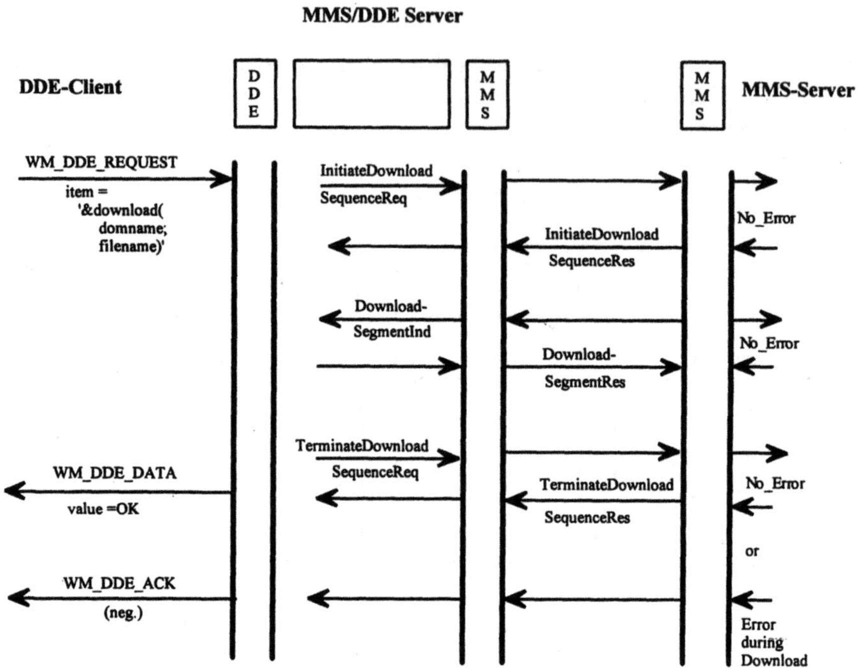

Figure 8.24: Downloading a domain

8.5 Example: calling MMS in an Excel spreadsheet

Spreadsheets are intensively used in offices to perform financial analyses. Increasingly, there is a demand for spreadsheets in other environments, for example on the shop floor. In the ESPRIT CCE-CNMA project [CCE 95], an MMS/DDE server has been developed and tested in various pilots in conjunction with different Windows applications. This section explains how Excel can get access to manufacturing devices through the MMS/DDE server.

The server can be accessed in three different ways from an Excel spreadsheet:

1. the EXECUTE macro
2. the REQUEST macro and
3. conventional formulae

The two macros are used in a similar way, except that MMS services must be prefixed by the "&" character in the case of the REQUEST macro:

=EXECUTE(channelno;"&identify()")
=REQUEST(channelno;"&identify()")

A formula has the following typical syntax: `ddemms'!'PLC1'!'var1'.

This formula establishes a hot link between Excel and the MMS/DDE server. It is composed of three parts:

1. The *application* parameter `ddemms!' is the name of the MMS/DDE server in the Windows environment.
2. The *topic* parameter `PLC1' is the name of the connection between the PC and the PLC1.
3. The *item* `var1' is the name of the variable in the PLC.

The result of the call appears in the cell containing the formula.

The following macro downloads the code of a program, creates a program invocation and starts it from Excel:

```
channel=INITIATE("ddemms", "s5server")
=REQUEST(channel;"&download(DOM_CONTROL,C:\CONTROL.DOM)")
=REQUEST(channel;"&pi_create(PI_CONTROL,DOM_CONTROL)")
=REQUEST(channel;"&pi_start(PI_CONTROL)")
=TERMINATE(channel)
```

The code of the domain is stored in a file located in the C: directory of the PC.

8.6 Summary

This chapter presented a simple interface to MMS services for the Microsoft Windows environment. At the heart of this operating system is a protocol called DDE (Dynamic Data Exchange) that allows Windows applications to communicate. DDE defines a number of communication links that can be established between a client and a server. These are the cold, hot and warm links. The MMS/DDE server is a program that allows Windows applications to access MMS services and thus to communicate with shop floor devices. The chapter presented the basics of DDE, the DDE interface to MMS services and the operation of the MMS/DDE server. An example was given in which an Excel spreadsheet is used to access data in manufacturing devices. With this mechanism it is possible to build new applications that reuse existing packages and tool kits of the Windows environment.

9. Other MMS Objects and Services

Chapter 4 introduced a number of MMS objects that are found in all existing MMS applications. Due to their generality, these objects are always implemented in manufacturing devices. The MMS Standard however defines additional objects and services that are sometimes implemented. These objects and services are concerned with operator communication, semaphores, events and journals. The purpose of this chapter is to briefly introduce these objects and services.

9.1 Operator stations

Terminals are widely used in industrial applications for communication with operators. In manufacturing, operators working on an assembly line get messages from production controllers instructing them of which operation must be performed on given parts. In process control, terminals are used to display the current state of process variables.

MMS provides a minimum functionality for communication with input-output devices. This functionality is sufficiently general to be applicable to a wide range of device types such as bar code readers, printers, terminals, displays, etc. For more sophisticated applications, the user can rely on protocols such as the OSI Virtual Terminal Protocol (VTP) or its equivalent in the Internet Protocol Suite, namely Telnet.

MMS defines the **Operator Station object** to cover three types of devices:

1. Entry devices, for example bar code readers
2. Display devices, for example displays, screens
3. Entry-display systems such as terminals composed of a display and keyboard.

MMS Operator Stations cannot be created or deleted as they are assumed to represent physical devices. Two services are defined on these objects:

1. **Input** allows a client to get a character string from an entry device. A prompt can be displayed before the string is read. This string can for example represent a question asked to the operator. The client can optionally specify a maximum waiting time to obtain a response from the operator.
2. **Output** is used to write a string to an output device.

Despite their simplicity, operator station services are rarely implemented in industrial applications.

9.2 Semaphores

In industrial applications, there is sometimes the need to synchronise the actions performed by multiple programs on a given resource. For example, when a server is accessed by multiple clients concurrently, it may be necessary to ensure that a single client at a time is modifying a given variable or set of variables. MMS semaphores are designed to solve such problems in MMS applications.

An MMS semaphore is composed of a set of tokens that can be allocated by a server to clients. When a token cannot be allocated, the client request is placed in a queue of pending requests. Once the token is released, it is allocated to the first client in the waiting queue.

MMS defines two classes of semaphores: **token** semaphores are composed of anonymous tokens while **pool** semaphores have named tokens that can be identified in client requests.

Two objects are defined in the standard:

1. A **Semaphore Entry** is a record that describes a request for a given semaphore. The request may be either in control of the semaphore, it may be pending or hung.
2. A **Semaphore object** describes a semaphore in terms of the tokens associated to it and the two queues representing the owners and requesters of the semaphore.

The following services are defined on MMS Semaphores:

- **TakeControl** is used to request a token from a semaphore. The client can specify the name of a token in the case of a Pool Semaphore. In addition, a priority can be specified for the request as well as an indication of the time the client is ready to wait for the semaphore or the time he is ready to keep control of it.
- **RelinquishControl** allows the client to release a previously granted semaphore.
- **DefineSemaphore** may be used to create a Token Semaphore in the remote user.
- **DeleteSemaphore** is the reverse operation allowing the removal of a Token Semaphore from a server.

- **ReportSemaphoreStatus** is defined to obtain the attributes of a semaphore, whatever its type. These attributes reflect the number of hung and owned tokens as well as the total number of tokens defined for the semaphore.
- **ReportPoolSemaphoreStatus** is a special service to obtain the name and the state of the tokens controlled by a Pool Semaphore.
- **ReportSemaphoreEntryStatus** may be used to obtain the detailed status of the semaphore entries associated to a semaphore.

A comparison of MMS semaphores with classical semaphores can be found in [Castori 95a].

9.3 Events

Events represent significant state changes in a system. They are used to notify normal as well as exceptional state changes. In a manufacturing environment for example, events are used to signal the completion of a manufacturing operation, the fact that a tool has broken, the arrival of a conveyor at a given destination, etc. Primitive forms of events exist in MMS in the form of Unsolicited Status and Information Reports. As shown in chapter 4, the conditions in which these reports are sent is not remotely configurable in MMS but is predefined. MMS events provide the functionality to do this remote configuration, not only of the conditions in which notifications are sent but also of actions to execute when events occur.

MMS distinguishes two categories of events: **Network-triggered events** represent events that can be explicitly triggered by a client using a dedicated MMS service. **Monitored events** represent on the other hand a state change of a VMD that is detected by this VMD through periodic monitoring.

The Standard defines a powerful functionality to handle events. Three objects are introduced that model different aspects of this functionality. The **Event Condition (EC)** object is a descriptor indicating the conditions that represent a given event. An **Event Action (EA)** is an MMS confirmed service that can be executed in the server when a specific event occurs. An **Event Enrollment (EE)** represents a subscription from a client to be notified when a given event occurs. The EE references an EC and optionally an EA.

The following services are defined on Event Conditions:

- **DefineEventCondition** creates an EC object and initialises this object with parameters provided by the client.
- **DeleteEventCondition** removes the EC object from the server.
- **GetEventConditionAttributes** provides the client with the attributes of the EC object.
- **ReportEventConditionStatus** provides the client with attributes that are dynamic in nature, for example the state of the object.
- **AlterEventConditionMonitoring** is used to turn on or off the periodic evaluation of the condition defining events. The user can also modify the period and priority of the evaluation process.

For the Event Action, the following services are defined:

- **DefineEventAction** creates an EA object and initialises this object with parameters provided by the client.
- **DeleteEventAction** removes the EA object.
- **GetEventActionAttributes** provides the client with the attributes of the EA object.
- **ReportEventActionStatus** provides the client with attributes that are dynamic in nature, for example the state of the object.

Event Enrollment objects can be manipulated with the following MMS services:

- **DefineEventEnrollment** creates an EE object and initialises this object with parameters provided by the client.
- **DeleteEventEnrollment** removes the EC object from the server.
- **GetEventEnrollmentAttributes** provides the client with the attributes of the EE object.
- **ReportEventEnrollmentStatus** provides the client with attributes that are dynamic in nature, for example the state of the object.
- **AlterEventEnrollment** allows the client to modify some of the attributes of the EE.

The basic services of MMS event management are:

- **TriggerEvent** to signal an event from an MMS user to one or multiple other MMS users through a server.
- **EventNotification** to notify a client of an event he has subscribed to. This notification indicates the present state of the EC, the time at which the event occurred and optionally the result of the action associated to this event.
- **AcknowledgeEventNotification** for a client to notify a server that he has well received a notification from this server.

More details on the implementation of an MMS event manager can be found in [Richner 93] and [Castori 95b].

9.4 Journals

Journals are widely used in industrial applications to record significant data and events. MMS Journals allow to chronologically record data and events of interest to an application. They can be explicitly created, deleted and reinitialised. Access to journals is possible by specifying filters concerning the start entry, the stop entry or the variables of interest to the client.

MMS considers three types of entries:

1. **Annotations** are character strings whose format is user defined.
2. **Event-data** describe significant events in terms of event condition names and variables associated to these events.
3. **Data**, contain the value of one or multiple variables.

MMS introduces two objects related to journals:

1. **A Journal Entry** is a record that can contain one of the three types of entries considered by MMS. It is the basic access unit of a journal.
2. **A Journal object** is a chronologically ordered collection of Journal Entries.

The following services are defined on Journals:

- **CreateJournal**: creation of a Journal object with specific attributes
- **DeleteJournal**: deletion of a Journal and all entries contained therein.
- **InitializeJournal**: Removal of all or some journal entries of a given journal.
- **WriteJournal**: Storage of one or more journal entries into a given journal.
- **ReadJournal**: Retrieval of one or more journal entries from a given journal. The client can specify a range of entries he is interested in. The start and end of the range are specified either by a time or an entry identifier. In addition, the client can include a filter that describes the variables he is interested in.
- **ReportJournalStatus**: retrieval of two attributes associated to a given journal, namely the number of entries and the MMS deletable attribute.

9.5 Summary

MMS defines objects and services that are applicable in a wide range of industrial applications. The majority of these applications rely on a small subset of MMS services for manipulation of domains, programs and variables. Chapter 4 introduced the MMS services for these objects. The scope of MMS is larger than what is currently used. In this chapter, objects and services related to Operator Stations, Semaphores, Events and Journals were briefly presented. Operator Stations provide the means to communicate with terminals, printers, bar code readers, etc. Semaphores allow for the synchronisation of applications. Events provide the means to notify MMS users of significant changes and to specify the conditions in which these notifications are sent. Journals are used to record events and significant data. In spite of their simplicity, these additional objects are rarely found in industrial devices. One of the reasons is that they are of interest to a small subset of applications.

10. Conclusion

10.1 What has been discussed

In Chapter 1, we described the evolution of the CNMA project, and the motivation for producing this book. Chapter 2 described the rationale for the introduction of MMS, whilst chapter 3 introduced the basic concepts that underlie MMS, the modelling approach and the standards related to MMS. Chapter 4 presented the most important MMS objects and services that are implemented by the majority of device manufacturers. Chapter 5 analysed the use of MMS in three types of real manufacturing devices. Chapter 6 described the use of MMS in a real manufacturing application. Chapter 7 presented MMSI, the application programming interface to MMS. Chapter 8 described a DDE interface to MMS that is available in the Windows environment. Chapter 9 introduced other MMS objects and services which are sometimes implemented.

In this last chapter, we concentrate on the importance of MMS, its multi-sectorial applicability, the real benefits to the user, interoperability, where to obtain more information on CNMA and MMS and, finally, what to expect of the future.

10.2 Why is MMS important ?

For a long time, the wide deployment of networks in industrial applications has been hampered by the absence of a standard communication protocol that would enable interworking between industrial devices of different manufacturers, of different types and ages. The Manufacturing Message Specification (MMS) is a standard, vendor independent application protocol that was designed for this purpose. It allows the user to concentrate on his application and not on communication aspects. The specification, how applications may communicate, can be made on a high level of abstraction. Network details and internals of vendor systems are hidden by the Virtual Manufacturing Device (VMD). It offers a wide range of services for remote access to variables, for transfer of programs and data to and from industrial devices and for remote control of programs.

10.3 Where is MMS being applied ?

It is generally recognised among industrialists that MMS provides an important milestone in the development of open systems for manufacturing, and represents a powerful tool for effective management of manufacturing processes at all levels of the factory. Significant testimonies to the acceptance of MMS can be demonstrated at many industrial sites throughout the world, and in particular Europe where currently, growth is greatest. While the main MAP/MMS installations are at automotive assembly plants, involving hundreds of computers and controllers connected to multiple MAP networks, MMS is now being applied in a very diverse range of industries such as aerospace, airports, breweries, electronics, highways, mining, power generation plants, petrochemical plants, steel plants and waste treatment.

The CNMA consortium have, over many years developed and validated MMS solutions for numerous European companies, including Aerospatiale, Aeritalia, BMW, British Aerospace, EFACEC, Renault, Magneti Marelli and Mercedes-Benz.

10.4 What is MMS used for ?

The concepts of MMS mean that it can have very wide applicability, at all levels of an organisation but some of the more typical applications of MMS include the following:

- Distributed Numerical Control Systems (DNC).
- Flexible Manufacturing Systems (FMS).
- Robot Assembly Cells and Lines.
- Manufacturing Cells and Lines.
- Process Control Systems.
- Machine Data Collection Systems.
- Production Data Collection Systems.
- Facility Monitoring and Control Systems (FMCS).
- Production Information and Control Systems (PICS).

Most organisations are likely to have more than one of the applications listed above. The real benefits of using MMS are not always apparent when considering an individual application of MMS. As will be discussed in the next section, many of the major business benefits of using MMS are only realised when a standardised approach is taken, looking at all opportunities within an organisation.

10.5 What are the main benefits of using MMS ?

In today's climate, it is very difficult to justify investment in new technology, the emphasis throughout industry is to make better use of the technology that exists and learn how to apply it better. Invariably, this means to standardise on solutions and provide improved integration across an enterprise. Within this scenario, there are two perspectives to consider: the user's and the vendor's.

Clearly, a vendor of MMS solutions has an interest in promoting the use of MMS because it provides the opportunity for doing more business. However, perhaps what is not so apparent is that standard solutions can provide vendors with other benefits, some of which are as follows:

- Rationalised development work.
- Improved products resulting from stable solutions.
- Simplified marketing.
- Access to multi-vendor networks and systems.
- Easier access to world markets.
- Easier to become a preferred equipment supplier.

MMS has evolved as a result of user demands, including those of the manufacturing application developer and system integrator. The benefits that could be expected by these users will vary by degree according to the type of organisation where it will be applied, the applications it is intended for and the choice of vendor solutions available on the market. The following list includes many of the benefits a user might expect:

- Better communication between manufacturing devices.
- It is a generic specification for *all* industrial automation applications.
- It is an accepted standard and can therefore be used by anyone.
- It is expandable for specific applications.
- It allows interoperability between equipment from different vendors.
- Application programs will require less knowledge of different devices and will become more portable.
- Decreased cost of application software development.
- Innovative products will come to market through the use of a single standard.
- The standards structure allows for future growth.
- MMS can be used across existing network technologies.
- Selection of best equipment for application.
- Computer/equipment vendor independence.
- Lower systems integration and support costs.
- One-time training only.
- Flexibility in additions and changes.

To consider the needs and interests from one's own perspective is only natural, but through the experience of the CCE-CNMA project both users and vendors have shown mutual interest in seeing standardised solutions being developed and made available in the market.

The result is lower systems costs

10.6 What about interoperability ?

The experience of the CCE-CNMA consortium, in developing and validating solutions in many industrial applications has demonstrated that to achieve interoperability between multi-vendor devices it is necessary for each vendor to apply a very formal test and validation procedure, as described below:

1. Each new device offering MMS services should be subject to conformance test against internationally accepted independent test tools (all services claimed for each type of device). Regressive testing should be an integral part of this cycle.
2. Each conformant device should then undergo interoperability testing against other fully validated devices available on the market. Apart from verifying that basic static interoperability can exist between different devices, it is necessary to simulate a range of typical industrial test scenarios for which the devices are intended to be used. This is to verify that unexpected conditions are dealt with correctly, and that the device yields adequate loading and performance figures.

By applying the procedure described above this will help to eliminate most of the problems normally expected when devices from different vendors are required to communicate with each other in a real industrial application. Conformance testing alone will not guarantee interoperability whereas, in practice, well designed interoperability tests do give more effective results.

10.7 Where can I obtain more information ?

The CCE-CNMA consortium actively supports open systems. During this phase and previous phases of the project several documents have been produced and made available in the public domain with the intention of helping the user develop a better understanding of open systems. Currently, these include:

1. CNMA Implementation Guide - a specification of the CNMA building blocks [CNMA 93].
2. 'A Guide to the CIM networking products of the CNMA partners' - describes the products available from vendors of the consortium, including Bull, Olivetti, Robotiker, Siemens, Silicomp, and SNI [CNMA 94].

These documents, and many more may be obtained from an information server using World Wide Web, details of which are described in Appendix I. Alternatively, these documents, and several end of contract reports for previous phases of the CNMA project (EP955, EP2617 and EP5104) may be obtained upon request from the IT Information Desk, CEC, Brussels.

The European MAP/TOP Users Group, EMUG have compiled a very useful reference guide [EMUG 94] which contains a comprehensive catalogue of MMS related data, such as MAP/MMS products and installations, conformance test centres, MAP/TOP User Groups and contact details.

10.8 So, what of the future ?

One of the problems when using MMS is the selection of appropriate objects and services to model such things as pallets, tools, conveyors etc. which are application specific objects. We discussed the problem for the transport system of a flexible manufacturing system and proposed a solution based on common MMS services. This problem is general for industrial applications and a standard solution is expected by users. The definition of standard mappings of application specific objects to MMS objects has been started with the MMS Companion Standards. This effort should continue with a goal to ensure a single definition that can be implemented by all vendors. Without this, the problem of heterogeneity, initially found by users at the communication level will stay at the application level with different representations of the same application objects.

Although MMS removes many of the problems associated with communications between manufacturing control devices, it is clear that popularisation of this protocol will also need the support of various enabling technologies. We have already discussed how MMSI and DDE can ease the problems with interfacing to MMS, the latter being one of the results from the CCE-CNMA project. But other technologies are emerging, such as OLE and CORBA which are likely to be influential in standardising the object oriented view of manufacturing.

The current phase of CCE-CNMA has looked at the *how* and *why* applications are developed, building on the successful work of CNMA and produced CCE: an integration platform for distributed manufacturing applications. The results of this project, and a survey of advanced computing technologies are published in a separate report, Volume 1 of this series [CCE 95].

In conclusion, we hope that you have been given an insight into the practical applications of MMS, something that was unfortunately lacking in published literature to date, and with this awareness and familiarity with the service go on to understand the significant benefits to be gained within your own industrial environment. The technical concepts described by MMS are good, and realisation of this fact will undoubtedly result in wider market opportunities for both user and vendor.

Bibliography

Throughout this book a number of references have been made to external publications, including books, articles, journals, proceedings, and standards the details of which are listed below in alphabetical order.

[Brockschmidt 93] K.Brockschmidt, "Network DDE in Windows for Workgroups 3.1 Bridges Programming Between PCs", *Microsoft System Journal*, January, 1993 (English edition), March/April 1993 (German edition).

[Castori 95a] P. Castori, "Semaphores revisited with MMS", to appear in *ACM Operating Systems Review*, July 1995.

[Castori 95b] P. Castori, "Performance Evaluation of an MMS Event Manager", *Technical Report EPFL-DI-LIT*, No. 93-203, also available in French in *Actes du Congrès CFIP'95*, Rennes, France, 1995.

[CCE 95] CCE-CNMA Consortium, "*CCE: An Integration Platform for Distributed Manufacturing Applications*", Springer-Verlag, 1995.

[CNMA 93] ESPRIT Project 7096, CCE-CNMA Consortium, "*CNMA Implementation Guide* ", Revision 6.0, November 1993.

[CNMA 94] ESPRIT Project 7096, CCE-CNMA Consortium, "*A Guide to the CIM networking products of the CNMA partners* ", January 1994.

[DecPle 93] J.-D. Decotignie, P. Pleinevaux, "A survey on industrial communication networks", *Ann. Télécommunications*, Vol. 48, No 9-10, 1993, pp 435-448.

[Digitalk 92] Digitalk Corporation, "Smalltalk/V 2.0 Win - Guide to Programming", 1992.

[EMUG 94] EMUG, "MAP/MMS Product and Installation Directory", 3rd edition, April, 1994.

[IROFA 92] IROFA, FAIS CellNet Specification, 1992.

[ISO/IEC 9506-1] ISO/IEC 9506-1, Industrial automation systems - Manufacturing Message Specification - Part 1: Services Definition, 1990.

[ISO/IEC 9506-2] ISO/IEC 9506-2, Industrial automation systems - Manufacturing Message Specification - Part 2: Protocol specification, 1990.

[ISO/IEC 9506-3] ISO/IEC 9506-3, Industrial automation systems - Manufacturing Message Specification - Part 3: MMS Companion Standard for Robotics, 1991.

[ISO/IEC 9506-4] ISO/IEC 9506-4, Industrial automation systems - Manufacturing Message Specification - Part 4: NC Companion Standard for MMS, 1991.

[ISO/IEC 9506-5] ISO/IEC 9506-5, Industrial automation systems - Manufacturing Message Specification - Part 5: Companion Standard for Programmable Controllers.

[ISO/IEC 9506-6] ISO/IEC 9506-6, Industrial automation systems - Manufacturing Message Specification - Part 6: Companion Standard for Process Control.

[Lotus 93] Lotus Corporation, Lotus 1-2-3 for Windows, Handbook, 1993.

[MAP 88] General Motors, Manufacturing Automation Protocol, Version 3.0, August 1988.

[MAP 93] World Federation of MAP/TOP Users Groups, Manufacturing Automation Protocol, Version 3.0, 1993 Release.

[Microsoft 92a] Microsoft Corporation, "Microsoft Windows Software Development Kit Version 3.1 - Reference Volume 1 to 4", 1992.

[Microsoft 92b] Microsoft Corporation, "Microsoft Windows Software Development Kit Version 3.1 - Guide to Programming", 1992.

[Microsoft 92c] Microsoft Corporation, "Microsoft Word for Windows 2.0", Handbook, 1992.

[Microsoft93a] Microsoft Corporation, "Microsoft Visual Basic 2.0 - Guide to Programming", 1993.

[Microsoft 93b] Microsoft Corporation, "Microsoft Excel 4.0", Handbook, 1993.

[Nussbaumer 91] H. Nussbaumer, *Téléinformatique IV*, Presses Polytechniques et Universitaires Romandes, 1991.

[Petzold 90] C. Petzold, "Programming Windows", 2nd edition, Microsoft Press, 1990.

[Pleinevaux 94] P. Pleinevaux, "An Analysis of the MMS Object Model", *IEEE Trans. on Industrial Electronics*, Vol. 41, No. 4, June 1994, pp 265-268.

[Richner 93] T. Richner, P. Castori, P. Pleinevaux, "Architecture of an MMS Event Manager", *Proc. of COMPEURO'93*, Paris, May 1993, pp 435-444.

[Siemens 92] Siemens Corporation, "TF/DDE-Manager", Handbook, 1992.

[Snyder 93] A. Snyder, "The Essence of Objects: Concepts and Terms", *IEEE Software*, January 1993.

[Valenzano 92] A. Valenzano, C. Demartini, L. Ciminiera, *MAP and TOP Communications*, Addison-Wesley, 1992.

[Wonderware 91] Wonderware Software Development Corporation, "NetDDE", Handbook, 1991.

Glossary of Terms

ACSE	Application Control Service Elements
AMIG	Australian MAP/TOP Interest Group
ANSA	Advanced Networked Systems Architecture
AOW	Asian OSI Workshop
API	Application Programming Interface
CCE	CIME Computing Environment
CIM	Computer Integrated Manufacturing
CIME	Computer Integrated Manufacturing & Engineering
CLNP	Connectionless Network Protocol
CMIP	Common Management Information Protocol
CNC	Computer Numerical Control
CNMA	Communications Network for Manufacturing Applications
CSMA/CD	Carrier Sense Multiple Access with Collision Detection
DDE	Dynamic Data Exchange
EEMIG	East European MAP/TOP Interest Group
EG	Expert Group
EMUG	European MAP/TOP Users Group
EPA	Enhanced Performance Architecture
EWOS	European Workshop on Open Systems
FAIS	Factory Automation Interconnection System
FDDI	Fibre Distributed Data Interface
FMS	Fieldbus Message Specification
FO	Fiber Optics
FTAM	File Transfer, Access and Management
GOSIP	US Government OSI Profile

IEC	International Electrotechnical Committee
IEEE	Institute of Electrical and Electronics Engineers
IROFA	International Robotics and Factory Automation Centre
ISO	International Standard Organisation
JMUG	Japanese MAP Users Group
KMIG	Korean MAP Interest Group
LAN	Local Area Network
MAP	Manufacturing Automation Protocol
MMS	Manufacturing Message Specification
MMSI	MMS Interface
NAMTUG	North American MAP/TOP Users Group
NC	Numerical Controller
NIK	Network Integration Kit
OIW	OSI Implementors Workshop
OSI	Open Systems Interconnection
PLC	Programmable Logic Controller
RC	Robot Controller
RDA	Remote Database Access
RMD	Real Manufacturing Device
SIG	Special Interest Group
TOP	Technical and Office Protocol
TP4	ISO Transport class 4 protocol
VMD	Virtual Manufacturing Device
VTP	Virtual Terminal Protocol
WAN	Wide Area Network
WFMTUG	World Federation of MAP/TOP Users Groups

Appendix 1. The MMS and CNMA Information Servers

The MMS Information Server is a WWW (World Wide Web) server that provides a number of different categories of information related to the MMS protocol. At the time of writing, the server contains:

- A bibliography of books, tutorials, research papers, technical reports concerned with MMS.
- A list of MMS products: MMS provider implementations for PCs, UNIX machines, embedded systems, manufacturing devices, application enablers, etc.
- Introductions to MMS, in the form of Frequently Asked Questions (FAQ) or more detailed presentations.
- A directory of persons interested in the MMS protocol - the MMS Interest Group - that points to the WWW pages of these persons.
- A directory of WWW servers related to manufacturing.

The MMS Information Server is accessed everyday by people from everywhere in the world. It was founded in November 1993, first as an FTP server and then turned into its present form when WWW became popular in 1994.

To access this server, you have to use one of the WWW browsers, for example Mosaic or Netscape, and open the following URL:

http://litwww.epfl.ch/~mms/mms.html

This action will directly bring you to the home page of this server. If you have problems or suggestions to make, you can contact the team that maintains this server by sending a mail to the following address:

mms@di.epfl.ch

A companion information server provides information on the CNMA and CCE-CNMA projects. This server contains a presentation of the project, a catalogue of products that are the result of these projects and a number of reports. To access the CNMA information server, open the following URL:

http://litwww.epfl.ch/~ppvx/CCE-CNMA.html

Appendix 2. MMSI Context Sensitive Functions

A2.1 Introduction

Context sensitive functions are the core of the MMSI application programming interface as they give access to the MMS services.

This appendix lists for each MMS functional unit, the name of the context sensitive functions and a short description is given of the function. For a detailed description of these functions, the reader is referred to the MAP 3.0/1993 supplement which is the latest specification of the MMSI interface.

A2.2 Context management

Context Management functions govern the basic operations to setup a communication environment for local Application Entity, establish a connection with a remote Application Entity, manage this environment and then close communication and clear this environment.

A description of the services available for Context Management is provided in Table A2.1.

Context Management	
Function Name	**Description**
mm_vactivation()	VMD Activation: Local service to set-up a local VMD to act as server.
mm_vdeactivation()	VMD Deactivation: Local service to deactivate a local VMD.
mm_aeactivation()	AE Activation: Local service to set-up and activate a local AE.
mm_aedeactivation()	AE Deactivation: Local service to deactivate a local AE when all connections are closed.
mm_connect()	Connect: High level service to establish a connection negotiated between a local and a remote AE.
mm_listen()	Listen: Local service to prepare a local AE to accept and receive connection indications.
mm_slisten()	Stop Listen: Local service to cease a local AE listening connection indications.
mm_answer()	Answer: Response to a connection indication received by mm_ireceive().
mm_ireceive()	Indication Receive: Local service to receive an indication from the indication queue.
mm_cnclude()	Conclude: Request to close properly a connection with the remote confirmation.
mm_coresponse()	Conclude Response: Response to A Conclude Indication.
mm_abort()	Abort: Request to abort a connection without waiting for completion of outstanding confirmed services.

Table A2.1: Services available for Context Management functions

A2.3 VMD support

VMD support functions provide services to obtain information about a remote VMD. The types of functions supported are:

- Identification of the VMD
- Status of the VMD
- List of the MMS objects contained in the VMD
- List of capabilities of the VMD.

The VMD can spontaneously send a notification indicating a status change.

VMD Support	
Function Name	**Description**
mm_Identify()	Identify: Confirmed service to request a server to provide the identification of the VMD.
mm_viresponse()	Identify Response: Response to an Identify Indication.
mm_status()	Status: Confirmed service to request a server to provide the status of a VMD.
mm_sresponse()	Status Response: Response to a Status Indication.
mm_ustatus()	Unsolicited Status: Unconfirmed service sent by a server to a client to report the status of the VMD without being solicited.
mm_gnlist()	Get Name List: Confirmed service to request a server to provide the list of the MMS objects of the VMD.
mm_nlresponse()	Get Name List Response: Response to a GetNameList Indication with the list of the MMS objects of the VMD.
mm_gclist()	Get Capability List: Confirmed service to request a server to provide the list of capabilities of the VMD.
mm_gcresponse()	GetCapabilityList Response: Response to a GetCapabilityList Indication with the list of capabilities of the VMD.

Table A2.2: Services available for VMD Support functions

A2.4. Domain management

Domain Management provides services to:

- Download to a Domain
- Upload a Domain
- Obtain the Attributes of a Domain
- Download/Upload a Domain requested from a server
- Download/Upload a Domain requested of a server

A description of the services available to support these functions can be found in Tables A2.3 to A2.7.

Domain Downloading	
Function Name	**Description**
mm_idsequence()	Initiate Download Sequence: Confirmed service to request a server to initiate a domain download sequence
mm_idlresponse()	Initiate Download Sequence Response: Response to an InitiateDownloadSequence Indication.
mm_dsegment()	Download Segment: Confirmed service sent by a server to request a client to download a segment into the server domain.
mm_dsresponse()	Download Segment Response: Response to a DownloadSegment Indication. The client responds with a data segment.
mm_tdsequence()	Terminate Download Sequence: Confirmed service sent by a server to terminate a domain download sequence.
mm_tdresponse()	Terminate Download Sequence Response: Response sent by a client to a Terminate-DownloadSequence Indication to confirm the download sequence is terminated.
mm_fdownload()	File Download: High level service sent by a client or a third party to perform a complete sequence of domain download: *mm_idsequence()* *mm_dsresponse(),..., mm_dsresponse()* *mm_tdresponse()*

Table A2.3: Services available for Domain Downloading.

Domain Uploading	
Function Name	**Description**
mm_iusequence()	Initiate Upload Sequence: Confirmed service to request a server to initiate a domain upload sequence.
mm_iulresponse()	InitiateUploadSequence Response: Response to an Initiate Upload Sequence Indication.
mm_usegment()	Upload Segment: Confirmed service sent by a client to request a server to upload a domain segment.
mm_usresponse()	Upload Segment Response: Response to an UploadSegment Indication with a server domain segment.
mm_tusequence()	Terminate Upload Sequence: Confirmed service sent by a client to terminate a domain upload sequence.
mm_turesponse()	Terminate Download Sequence Response: Response to a TerminateUploadSequence Indication.
mm_fupload()	File Upload: High level service sent by a client or a third party to perform a sequence of domain uploading: *mm_iusequence()* *mm_usegment(),..., mm_usegment()* *mm_tusequence()*

Table A2.4: Services available for Domain Uploading.

Domain Support	
Function Name	**Description**
mm_gdattribute()	GetDomainAttributes: Confirmed service to request a server to provide the attributes of a specified domain.
mm_daresponse()	GetDomainAttribute Response: Response of a server to a GetDomainAttribute Indication.

Table A2.5: Services available to obtain the Attributes of a Domain.

Server requests for Domain Downloading/Uploading	
Function Name	**Description**
mm_rddownload()	RequestDomainDownload: Confirmed service sent by a server to request a client or a third party to download a domain of the server.
mm_ddresponse()	RequestDomainDownload Response: Response sent by the client or a third party to a RequestDomainDownload Indication when a download sequence is terminated.
mm_rdupload()	RequestDomainUpload: confirmed service sent by a server to request a client or a third party to upload a domain from the server.
mm_duresponse()	RequestDomainUpload Response: Response sent by a client or a third party to a RequestDomainUpload Indication when the upload sequence is terminated.

Table A2.6: Services available for Domain Download/Upload from a Server.

Services to request domain Downloading /Uploading	
Function Name	**Description**
mm_ldcontent()	LoadDomainContent: Confirmed service to ask a server to request a domain download from a file server.
mm_ldresponse()	LoadDomainContent Response: Response of a server to a LoadDomainContent Indication when the request domain download is terminated.
mm_sdcontent()	StoreDomainContent: Confirmed service to ask a server to request a domain upload to a file server
mm_sdresponse()	StoreDomainContent Response: Response of a server to a StoreDomainContent Indication when the requested domain upload is terminated

Table A2.7: Services available for Domain Download/Upload requested of a Server.

A2.5 Program invocation management

The Program Invocation Management functions provide the following services:

- Create a program invocation composed of one or more domains linked together to form an executable task and specify its attributes: name, list of domains, shareable, deletable, etc.
- Delete existing and deletable program invocations.
- Get the attributes of a program invocation.
- Start a program invocation with an initialization parameter.
- Stop a program invocation which is running.
- Resume a program invocation which is stopped.
- Reset a program invocation which is stopped and set it in the idle state.
- Kill a program invocation, whatever its state.
- Get the attributes of a program invocation.

All these services are described in Table A2.8.

Program Invocation Management	
Function Name	**Description**
mm_cpinvocation()	CreateProgramInvocation: Confirmed service to request a server to create a program invocation on one or many domains.
mm_cpresponse()	CreateProgramInvocation Response: Response to a CreateProgramIndication.
mm_pidelete()	DeleteProgramInvocation: Confirmed service to request a server to delete a program invocation.
mm_dpresponse()	DeleteProgramInvocation Response: Response to a DeleteProgram Indication
mm_strt()	Start: Confirmed service to request a server to start a program invocation.
mm_sttresponse()	Start Response: Response to a Start Indication.
mm_stop()	Stop: Confirmed service to request a server to stop a program invocation
mm_stpresponse()	Stop Response: Response to a Stop service Indication.
mm_resume()	Resume: Confirmed service to request a server to resume a program invocation.
mm_rsmresponse()	Resume Response: Response to a Resume Indication.
mm_rst()	Reset: Confirmed service to request a server to reset a program invocation.
mm_rsresponse()	Reset Response: Response to a Reset Indication
mm_kill()	Kill: Confirmed service to request a server to kill a program invocation.
mm_kresponse()	Kill Response: Response to a Kill Indication
mm_gpinvocation()	GetProgramInvocationAttribute: Confirmed service to request a server to return the attributes of a program invocation.
mm_piaresponse()	GetProgramInvocationAttribute Response: Response to a GetProgramInvocationAttribute Indication.

Table A2.8: Services available to Support Program Invocation Management.

A2.6 Variable access

The Variable Access functions provide the following services:

- Read the values of one or more variables maintained in the MMS server.
- Write the values of one or more variables maintained in the MMS server.
- Report the values of variables maintained in an MMS server without solicitation from the client.
- Get the attributes of one or more variables maintained in the MMS server.

All these services are described in Table A2.9.

Variable Access Functions	
Function Name	**Description**
mm_read()	Read: Confirmed service to read a list of variables maintained by a server.
mm_rresponse()	Read Response: Response to a Read Indication with the values of the requested variables.
mm_write()	Write: Confirmed service to request a server to write the values of a list of variables.
mm_wresponse()	Write Response: Response to a Write Indication with a confirmation of the request.
mm_rinformation()	Report Information: Unconfirmed service sent by a server to report without solicitation the values of a list of variables maintained by the server.
mm_gaattribute()	GetAccessAttribute: Confirmed service to get the attributes of a variable maintained by the server.
mm_gvresponse()	GetVariableAccessAttribute Response: Response to a Get Variable Attribute Indication.
mm_dlvariable()	Define Local Variable: Local service to define a local variable maintained by the MMSI Service Provider or the Automatic Responder.
mm_glvariable()	Get Local Variable: Local service to get the attributes of a local variable.
mm_lvdelete()	Delete Local Variable: Local service to delete a local variable.

Table A2.9: Services available to support Variable Access Management.

Research Reports ESPRIT

Area *Computer-Integrated Manufacturing and Engineering (CIME)*

Improving the Performance of Neutral File Data Transfers. Edited by R.J. Goult, P.A. Sherar. IX, 138 pages. 1990 (Project 322 CAD*I, CAD Interfaces, Vol. 6)

Advanced Modelling for CAD/CAM Systems. Edited by H. Grabowski, R. Anderl, M.J. Pratt. VI, 113 pages. 1991 (Project 322 CAD*I, Vol. 7)

IMPPACT Reference Model. Edited by W.F. Gielingh, A.K. Suhm. XII, 261 pages. 1993 (Project 2165 IMPPACT, Integrated Modelling of Products and Processes using Advanced Computer Technologies)

CIMOSA: Open System Architecture for CIM. Edited by ESPRIT Consortium AMICE. XI, 234 pages. 2nd, rev. and ext. edition 1993 (Project 688/5288 AMICE, A European CIM Architecture)

Vibration Control of Flexible Servo Mechanisms. Edited by J.-L. Faillot. VII, 206 pages, 1993 (Project 1561 SACODY, A High Performance Flexible Manufacturing System (FMS) Robot with On-Line Dynamic Compensation)

CCE: An Integration Platform for Distributed Manufacturing Applications. A Survey of Advanced Computing Technologies. Edited by ESPRIT Consortium CCE-CNMA. XII, 207 pages. 1995 (Project 7096 CCE-CNMA, CIME Computing Environment: Integrating CNMA, Vol. 1)

MMS: A Communication Language for Manufacturing. Edited by ESPRIT Consortium CCE-CNMA. XII, 183 pages. 1995 (Project 7096 CCE-CNMA, Vol. 2)

Subseries PDT (Product Data Technology)

CAD Geometry Data Exchange Using STEP. Edited by H.J. Helpenstein. XIV, 432 pages. 1993 (Project 2195 CADEX, CAD Geometry Data Exchange)

Neutral Interfaces in Design, Simulation, and Programming or Robotics. Edited by I. Bey et al. XV, 334 pages, 6 figs. 1994 (Project 2614/5105 NIRO, Neutral Interfaces for Robotics)

NEUTRABAS. A Neutral Product Definition Database for Large Multifunctional Systems. Edited by H. Nowacki. XII, 203 pages. 1995 (Project 2010 NEUTRABAS)

Springer-Verlag and the Environment

We at Springer-Verlag firmly believe that an international science publisher has a special obligation to the environment, and our corporate policies consistently reflect this conviction.

We also expect our business partners – paper mills, printers, packaging manufacturers, etc. – to commit themselves to using environmentally friendly materials and production processes.

The paper in this book is made from low- or no-chlorine pulp and is acid free, in conformance with international standards for paper permanency.

GPSR Compliance

The European Union's (EU) General Product Safety Regulation (GPSR) is a set of rules that requires consumer products to be safe and our obligations to ensure this.

If you have any concerns about our products, you can contact us on ProductSafety@springernature.com

In case Publisher is established outside the EU, the EU authorized representative is:

Springer Nature Customer Service Center GmbH
Europaplatz 3
69115 Heidelberg, Germany

Batch number: 10704713

Printed by Printforce, the Netherlands